The Fall of
the Wild,
the Rise of
the Zoo

OTHER BOOKS BY THE AUTHOR

*The Strenuous Decade: A Social and Intellectual Record
of the Nineteen-Thirties (edited with Daniel Aaron)*

The Politics of Schools: A Crisis in Self-Government

*Just Around the Corner: A Highly Selective History
of the Thirties*

Obstacle Course on Capitol Hill

White House Fever

The Riddle of the State Department

The Fall of the WILD

The Rise of the ZOO

Robert Bendiner

E. P. DUTTON ‖ NEW YORK

Published in the United States by Elsevier-Dutton Publishing Co., Inc.,
2 Park Avenue, New York, N.Y. 10016

Library of Congress Cataloging in Publication Data
Bendiner, Robert.
The fall of the wild, the rise of the zoo.
Bibliography: p. 152
Includes index.
1. Zoological gardens. I. Title.
QL76.B46 1981   590'.74'4   80-39704
ISBN: 0-525-10270-1

Published simultaneously in Canada by
Clarke, Irwin & Company Limited, Toronto and Vancouver

Designed by The Etheredges

10 9 8 7 6 5 4 3 2 1

First Edition

For Kas

Contents

Sixteen pages of illustrations follow page 56.

The Fall of
the Wild,
the Rise of
the Zoo

Introduction

At a dinner party not long ago I happened to remark in the course of some talk about endangered animals that I was working on a book about zoos. "Good, I'm sure you will be for closing them all down," said the lady on my left, and the sentiment was echoed around the table in various ways, as though no one with the slightest concern for the vanishing orangutan or the great bustard could possibly favor their being penned up in animal bastilles instead of being allowed to run free and multiply.

As it happens, I was enthusiastically in favor of the good zoos, of which there are a steadily growing number (the bad ones should be closed down tomorrow, if not this afternoon) but I was unable to explain why at the time because I knew it would take a whole book to do it. I am for zoos, as I hope to demonstrate, precisely because in what remains of the wild the orangutans and great bustards—as well as the Siberian tigers, one-horned rhinos, green turtles, and all too many other creatures—are not destined to run free and multiply at all but rather to sink into the zoological burial ground that already contains the last remnants of the dodo, the woolly mammoth, and the passenger pigeon. The great zoos in contrast—and in contrast to the cramped, smelly, boring little zoos of your childhood as well—are even now breeding more animal life than they take from forest, swamp, and plain, contributing more to the world's Siberian tiger population, for example, than do all the forests of Asia. And they exhibit that wildlife with a knowledge and feeling that can only revive in the people of an increasingly concrete urban culture a sense of wonder at the life they share on this planet, a kinship with it, and a strong desire to keep it from dropping into oblivion.

That there are such sentiments to be revived, or at least inspired, can hardly be doubted by anyone who has ever witnessed the visitors' appreciation of the wonderful World of Birds at New York's Bronx Zoo, or the exciting San Diego Wild Animal Park in California, or the spacious beauties of England's Whipsnade, or the imaginative exhibits of the Frankfurt Zoo in West Germany, to name at this point only a very few of the world's great animal institutions. Such witnesses would be even more impressed if they could go behind the scenes at these and many other excellent zoological parks, as I have had the privilege of doing.

Animal collections have still the enormous appeal that they have had through the centuries, going back through the menageries kept by the Medici popes to the apes and peacocks that King Solomon imported along with his ivory, gold, and silver. Even in a sports-mad country like ours, let it be noted, more people visit zoos in a year than are present at all the professional major league baseball and football games put together.*

That must mean something. But it is to these zoo buffs only secondarily that I address myself, because they are already convinced. My primary interest is in those environmentalists and animal lovers of all sorts who are emphatically *not* zoo buffs, in the hope that they may come to see how the good zoo, besides being endlessly fascinating, has an enormous potential for informing the public, for furthering scientific study, and, above all, for providing a safety net for species that, thanks mostly to man's incursions on their habitat, are probably not long for this planet.

---

*This startling comparison, first propounded, so far as I know, by Dr. William G. Conway, director of the New York Zoological Society, seems excessive on the face of it. But annual zoo and aquarium attendance comes to more than 100 million, whereas major league baseball, with the longest sports season, has attracted no more than 43.5 million fans in its best year and football less than one-third of that. What is more, adults outnumber children in the zoo attendance figures by three to one.

As a Practical Matter

If species can prove their worth through their contributions to agriculture, technology and other down-to-earth activities, they can stake a strong claim to survival space in a crowded world.

—NORMAN MYERS

Americans whose sympathies were all with the growing movement to protect the world's wildlife found themselves a few years ago in a trying position. An obscure three-inch fish known to the small brotherhood of ichthyologists as the snail darter was threatened with extinction by the building of a $120-million dam in Tennessee. Viewed superficially, the cost of rescuing the darter by abandoning the dam seemed somehow disproportional to its role in the scheme of things. But in a suit brought under the Endangered Species Act, the Supreme Court took the law as it found it and sided with the minnow. Work on the nearly completed dam was stopped cold.

It was an unprecedented victory for the conservationists but, many thought, a Pyrrhic one. They feared a widespread public feeling that environmentalism had been carried too far, and in fact Congress, quick to react, did modify the law. The change was not as drastic as some animal protectionists had anticipated, but talk in the Senate ran to phrases like "revulsion against environmental excesses" and such sentiments as that of Senator Jake Garn of Utah, who did not "give a damn if a fourteen-legged bug or the woundfin minnow lives or dies."*

The cause of wildlife protection may have been momentarily set back by the episode but time and reflection only bring credit to the snail darter's defenders. Let it be conceded at once that it is hardly possible to find for the darter and similarly unpretentious animals much of a role in man's evolving cultural patterns, such as one might reasonably cite in the case of the tiger or the wolf. Indeed, unless one's reverence for life is truly all-embracing, down to the tsetse fly and the Norway rat, the case for such creatures as the snail darter is left to rest on a hard-headed showing of utility.

As it happens, however, all animal life, like all plant life, *can* make that showing—either as fact or as potentiality. The truth is that we cannot know, once a species has been pushed into oblivion, or even casually allowed to drift toward it, what great advantage might have been discovered if, after a million years of evolving, it had been helped to survive for just one more decade.

Consider the possibility that at the turn of the century a threat of extinction had befallen an insect known to science as *Drosophila melanogaster* and to the rest of us as the fruit fly (more accurately designated the pomace or vinegar fly). Would the most ardent conservationist in the world, or even the most cool-headed scientist, have rushed to its defense on the off chance that it might some day be useful to man instead of merely a pest addicted to overripe fruit and vinegar? Not unless he had a morbid desire to become a target of national ridicule. Yet within a decade, T. H. Morgan, a young biologist, was to hit upon the tiny bug (twenty of its kind might make a good meal for a snail darter) as a likely subject for some investigations he was making into the nature of the gene as a physical entity in the workings of heredity.

What prompted Dr. Morgan's choice of the fly was that however much a nuisance *Drosophila* may have been in the kitchen, it was a star performer in its own bedroom, reproducing at a rate of about one generation a week. Crossbreeding, moreover, could be done in ideally small space and at ideally low cost. And to make the thing perfect, the insect's chromosomes were large and mutations common. In consequence, Morgan and his associates were able, by 1915, to prove and describe the operations of the genetic system in a book called *Mechanism of Men-*

*One of the hardships that animal life is subjected to is the nomenclature wished upon it by man. It is easier to work up ridicule than sympathy for the goitered gazelle, the marbled polecat, the pig-footed bandicoot, and many other creatures whose names do little for them on the banners of crusading conservationists.

delian Heredity. In his subsequent works the system was fully laid out, to become one of the pillars of modern biology. Dr. Morgan won a Nobel Prize and the so-called fruit fly won recognition, at least in select circles, for its contribution to humanity.

Comparably improbable as a benefactor to man was the rhesus macaque. Until recently this common monkey had been swarming over the cities and villages of India, fewer and fewer of the species caring to remain in the fast-shrinking forests when plunder was so available around humans. Dr. Walter Fiedler, director of the Schönbrunn Zoo, writes of the Indian War Ministry's complaint that rhesus monkeys were infesting its offices to steal official papers which they had learned could then be returned for fruit and candy, indicating no doubt a critical faculty on the part of the monkeys as well as a certain commercial shrewdness.

Yet even then the rhesus was being shipped to the United States at the rate of close to a quarter of a million a year in what would prove a highly successful effort to produce a polio vaccine. Some of the monkeys were used to test other vaccines as well—for smallpox, measles, and mumps. And that was by no means the limit of their contribution to human medicine. It had been found that the animal's blood contained what was to be called the Rh factor—Rh for rhesus. The study of that blood factor led to a system for sparing newborn babies the needless risk of a generally fatal blood disease caused by the combination of Rh positive and Rh negative parents.

Given the anatomical resemblances within the primate order, an evolving science of medicine has increasingly found uses for other monkeys. The cotton-top marmoset has proved to be a model for the study of colonic cancer, and the night monkey, a primitive New World species, has been useful in the study of certain antimalarial drugs. The vervet, or green monkey, and the crab-eating macaque can be used interchangeably with the rhesus for testing anti-polio vaccine, which is fortunate because the price of rhesus monkeys skyrocketed as their habitat decreased and the species declined in numbers. By the spring of 1978 the Indian government had understandably banned all further export.

Not all the work with these primates, it may be noted in passing, requires the kind of experimentation that would be fatal to the animals*; none of it, moreover, need, or ever should, involve inhumane treatment.

---

*Much scientific work on animals now involves the use of cell cultures and tissue, with painless testing devices, including computers, replacing to a great extent the sometimes inexcusably ruthless laboratory techniques of the past.

With the Rh factor, for example, what was required of a rhesus monkey was no more than what is required of a human who donates to a blood bank. It is also true, incidentally, that of the proportionally small number of monkeys used for medical purposes, a great many would have met less attractive fates at home. In Peru, for example, some 1.5 million squirrel monkeys were captured over a ten-year period to be exported for all purposes, including a huge sale as pets, and in the same decade some 7.5 million were reported to have been killed for food.

Nevertheless, successful efforts have since been made to substitute more populous species for the declining primates. At the Penrose Institute, connected with the Philadelphia Zoo (but forbidden to experiment with any of the zoo's live animals) the common woodchuck is being profitably studied for a hepatitis virus that is close, if not identical to the one afflicting humans. Dr. R. L. Snyder, who directs the program, points out that research on the woodchuck, common in wooded suburban areas, has made unnecessary the possibly alternative services of 180 scarce chimpanzees a year.

Primates, however, constitute only one per cent of all the animals used in scientific research. In medicine alone the honor roll of such species is long and growing annually, as the following examples suggest:

Although humans have looked down on the snake ever since the Garden of Eden, they have not failed in recent times to include its venom in their pharmacopoeias, and not just as an antidote to its own poison. Some venoms act as a styptic agent to stop bleeding, others are nonaddictive painkillers, particularly cobra venom, and in dilute form have been tried in the treatment of epilepsy. Venom derivatives are also being explored for possible use in treating asthma, arthritis, multiple sclerosis, and certain circulatory disorders.

ITEM: Apart from the fact that much of the plant life consumed by man depends on the bee for pollination—fruits and legumes especially —that insect has also contributed generously to the medicine cabinet. Some palliatives, if not cures, for arthritis contain bee venom; and the honey manufactured by bees is an ingredient of cough syrups, a high-energy item in prescribed diets, and a help in relieving the symptoms of hay fever.

ITEM: An unglamorous creature that has proved advantageous to the health of man is the blister beetle, which produces cantharidin, a substance that has been used to treat disorders of the urogenital tract. Likewise the electric eel, which Japanese medicine has found to yield an antidote for certain types of poisoning caused by insecticides.

ITEM: Two authorities on gastropods have reported that "specific glandular substances from the edible snail, garden snail . . . cause agglutination of certain bacteria, and therefore could be of therapeutic value against whooping cough, asthma, and other diseases."* And the blood of the horseshoe crab has been used to detect toxins in intravenous fluids.

ITEM: Smaller than the snail darter and about as close to extinction, the pupfish of Death Valley has managed to survive extremes of physical variation, including temperatures that range from freezing to well over 100 degrees Fahrenheit. The little fish's fantastic adaptability—it thrives in waterholes six times saltier than sea water—makes it invaluable for studies in evolutionary processes, and the direct application to human welfare promises to be even more specific. According to the National Science Foundation, conditions tolerated by the pupfish "tell us something about the creatures' extraordinary thermoregulatory system and kidney function—but not enough as yet . . . They can serve as useful biological models for future research on the human kidney—and on survival in a seemingly hostile environment."

ITEM: Sponges appear in depictions of bathrooms and bedsides going back to the Bronze Age. Dr. Ernst F. Kilian, an authority on the subject, writes that "in ancient times they were used not only to clean the body, as paintings on ancient vases show, but also to stop bleeding, as compresses for the cervix and boils, as pessaries, and also a type of gas mask as a protection against diseases."** Although these marine creatures have been largely but not entirely supplanted in the bathroom by synthetic articles, they have taken on a new medical potential with evidence that they may shed light on the cure of certain viral diseases.

ITEM: The manatee, or sea cow, appears to share with European royalty a blood that is slow to clot. It is considered a possibly useful

*They are also credited with such tonic effect that Napoleon's soldiers are said to have carried them, on orders, as emergency rations. Wellington's victorious men must have relied on the still more efficacious honey.
**Grzimek's Animal Life Encyclopedia, Vol. 1, p. 165.

model in future studies of hemophilia. Another animal thought to have usefulness in the study of blood clotting is, not unexpectedly, the vampire bat, whose saliva serves to inhibit the clotting that might otherwise interfere with its sanguine diet.

ITEM: Pharmacology researchers at Xavier University in Louisiana have found that alligator tongue oil produces a steroidal action in the human body that is claimed to give temporary relief for asthma, rheumatism, and arthritis.

ITEM: Earthscan, a media information unit serving the International Institute for Environment and Development, lists in addition to the animal "uses" mentioned here the medical contributions of certain birds: "The stormy petrel, the albatross, and other long-flying oceanic birds with highly developed heart and circulatory systems contribute [to] a better understanding of cardiomyopathy, an obstruction of blood outflow caused by overdevelopment of the heart muscle."

ITEM: Besides the marmoset, already mentioned, a number of animals have been drawn into the campaign against cancer. I was shown studies by Dr. Snyder and his colleagues at Penrose that linked hepatitis in the woodchuck not only with that viral disease in humans but also with liver cancer. At the San Diego Zoo, across the country from Penrose, studies made by the staff pathologist show a statistically high occurrence of tumors, many of them malignant, in marsupial animals such as the Tasmanian devil. Those so afflicted contribute to further studies in comparative pathology.

ITEM: Among the unexpected sources of help to modern medicine is the armadillo. This primitive creature turns out to be the only known animal besides humans that contracts leprosy, thereby holding out the hope that it will one day yield a cure, possibly a vaccine, for one of man's oldest and most unpleasant disorders. What makes the armadillo an especially good model for study is that it regularly produces four identical offspring at a time, thus offering an excellent opportunity for comparative observation and controlled experiments. The same animal, at least the nine-banded species, is of further medical interest, it happens, for an ability to hold its breath for as long as six minutes, presumably in order to lessen the amount of dust inhaled in the course of its fast and furious digging.

Given time and study, the life habits and anatomical wonders of many other animals can be expected to yield great rewards in knowledge bearing on human health. Research into hibernation by certain species of bears is expected to yield information on low-protein, low-fluid diets that may prove helpful to sufferers from kidney diseases. The lungfish may have something to teach man about the suspension of metabolic processes that could be useful in prolonged and delicate surgery.

In the same way, the vulture could shed light on correctives for food poisoning. The giant nerve cell of the squid may permit significant advances in the study of nervous physiology. It is supposed, though not yet established, that the sloth, an animal that spends its life upside down, might be instructive on such subjects as heart action, respiration, and the orientation of humans in outer space. And who knows what the geneticist might learn from the parthenogenesis of bees and certain lizards?

Long before the development of antibiotics but as late as World War I, maggots were used to treat long-standing ulcers or chronic bone infections. The larvae discharged allantoin, a substance that promoted the growth of healthy tissue. Luckily, no one ever had to be concerned about the possible extinction of maggots.

Far more than animals, plants are used to minister to man's infirmities. Even in this age of synthetics, approximately half the drugs used in Western medicine are based on substances originally found in nature, with the proportion much higher in Oriental medicine. The vast contribution of plants is beyond the scope of this book, but it is relevant here that a great many of these plants are dependent on animals, just as all animals are dependent, directly or indirectly, on plants.

Simply by feeding on them, animals crop the earth's plant life, keeping whole species from being starved or suffocated by their very number—and incidentally from blocking the flow of waters. Manatees for example, consume huge quantities of plant material daily; in consequence several Latin American countries a few years ago called on UNESCO to help get these sea cows into their plant-choked rivers in order to make them passable again for the fishing industry on which the countries depended. The hippopotamus serves the same purpose on some of the rivers of Africa.

This interdependence exists everywhere. Did the Great Plains make the American buffalo, or did the buffalo make the plains? Standing alone, neither proposition is true; taken together, both are.

The most common and obvious relationship between plant and animal, of course, is the latter's role in dispersing seeds. Tropical forests in particular, with almost no wind-pollinated trees, depend on animals for this function. Bats and birds distribute seeds, just as insects distribute pollen. Some animals pass large, hard-shelled seeds through their systems, discharging them in a condition softened enough by digestive acids to allow eventual germination. Other animals unwittingly carry seeds on their feet, tramping them into the ground as they go, or wittingly bury them for future consumption in the manner of squirrels sequestering acorns.

A dramatic instance of this exchange of services was noted in 1977 by Dr. Stanley A. Temple, a University of Wisconsin ornithologist. Writing in *Science*, Dr. Temple reported that on the island of Mauritius in the Indian Ocean, only thirteen specimens remain of the once-flourishing calvaria tree. All are more than three centuries old, none of these beautiful hardwoods having come into being, reportedly, since the killing of the last dodo in the seventeenth century. The calvaria's thick-hulled seeds, it seems, were worn down as they passed through the alimentary tract of that flightless bird and so made ready to sprout. The last hope for saving the tree species, according to Dr. Temple, lies in the successful import and substitution of turkeys for dodos in the seed-dispersal process.*

Thanks to a burgeoning appreciation of the environment, the best perceived "use" of wild creatures is the simple playing out of their ecological role in a natural chain that must be viewed as a whole, dependent for its integrity on the performance of each of its participants.

Although this concept has been helpful in recruiting and holding much needed support for animal conservation, it has no doubt been somewhat stretched and oversimplified. It is not quite true that the removal of a species necessarily means the collapse of an entire ecological web. Deplorable as it would be for other reasons, the disappearance of the whooping crane, if it were to occur, would no longer have any such impact. "The ecosystem won't even know when it's gone," says Earl Baysinger, formerly a top official of the Interior Department's Office of Endangered Species. It is too far gone already, he explains. So is the tiger, for that matter. There are simply too few of these great cats left to rest the case for preservation on their present niche in the ecology of

---

*Unless, of course, interested Mauritians undertake to soak the seeds in acid baths of their own devising and then plant them—a scheme that would clearly benefit future Mauritians rather than the generation doing the work.

India. Sufficient utilitarian reasons exist, as has been suggested—and intangible ones of even greater significance, to be discussed elsewhere —for saving these species without resorting to an argument of questionable validity.

Nevertheless, there *are* animals, many of them, whose ecological function is still so considerable that their swift elimination would in fact work powerful changes in the natural balance. Keeping those webs intact must by any calculation be regarded as a major animal "use." I recall asking Dr. Inge Poglayen, curator of mammals at the Arizona–Sonora Desert Museum, about the giant saguaro cactuses that set off that remarkable park in the desert west of Tucson. If any serious number of these giants—they stand as high as fifty feet—should be toppled by a rare and sudden deluge, she explained, there would be too little nesting area left for the Gila woodpeckers and owls. Very soon the insects and mice that they feed on would overrun the place, attracting predators from the surrounding areas, lured by the prospect of "delicious food on the table at all times." The nature of the area would be quickly and drastically altered, she said, adding almost as an afterthought, "Of course the cactus itself would not have been there at all if it had not been pollinated by bats in the first place."

Wherever one turns in nature, the story is the same. When ranchers and farmers of the American Southwest exterminated the red wolf as a pest, in came the more adaptable coyote to mate with the remnant wolf population and more than fill the gap. When Central Africans killed off thousands of hippos as crop-destroyers, they were soon to learn that their folly was to cost them a more serious dietary loss: the hippos' waste had been fertilizing the river algae which nourished the natives' favorite food, the talapia, a small fish now sadly depleted. In another aspect of existence, rural South Americans were to discover that when they had all but exterminated the caiman, more of their sheep than ever before died from the fatal attentions of the liver fluke, a parasitic flatworm. The crocodilians, it appeared (too late, as always) had performed the desirable function of eating great numbers of the giant snails that play host to that deadly fluke. Unwittingly, the ancient Egyptians did better by the ibis, which devoured great quantities of the snails that are host to the parasite responsible for schistosomiasis. They made it a god.

Such stunning ecological surprises are a fair measure of the ignorant nonchalance with which the world generally confronts the complexity of the animal world. When the Chinese twenty years ago opened war on sparrows, to cite another example, they had to learn that insects, vastly multiplied in consequence, could take a much bigger share of their crop than the birds had ever done. Similarly, the annual massacre of birds in Italy, including falcons and other birds of prey, has contributed heavily to an upsurge of vipers that, freed from the threat of predators, plague farmers and discourage Romans from spending weekends at their country villas.

Bats, which are well regarded chiefly by those engaged in the profitable guano trade, turn out to consume billions of insects in a night, especially destructive moths—a service barely realized, much less appreciated, by whole populations educated in bat lore solely by the creator of Dracula.

Even creatures that are more reasonably regarded as destructive turn out to be benefactors. When I trained long ago as an infantryman at Fort Rucker, Alabama (then Camp Rucker) I would occasionally wander into the nearby town of Enterprise. All I can recall of that metropolis now is that at its heart is a statue honoring the boll weevil, without whose fondness for the cotton plant the community would never have exchanged its dependence on that declining crop for the profitable exploitation of the peanut. In the same way a monument to the gull in the center of Salt Lake City marks the historic appreciation of Mormon farmers for the bird's service in devouring great swarms of crickets intent on eating the produce of their fields. If the Mormon settlers had destroyed the birds first, they might have thought at the moment that they were ahead of the game, but in the end they would have starved.

To take an extreme case, even the tsetse fly—which to be tolerated must certainly require a fanatical reverence for life—has its hard-nosed defenders. V. B. Wigglesworth, in *The Life of Insects*, calls it "the great savior of the African soil." What it saves the soil from is man, who "by ignorant procedures, or by ruthless exploitation for short-term gains," has wrecked great areas of the continent. He would wreck many more but for the fact that he dare not go where the land is occupied by the tsetse fly, which thereby "conserves the soil for the more enlightened cultivators of the future."

Comparably, as the water of the Everglades is drained off by developers, the alligator, using its sturdy snout, digs out a new " 'gator hole," which serves not only as a home for itself but as a haven for smaller aquatic species that follow it and would otherwise be doomed. Among the beneficiaries are millions of gambusia, tiny fishes that live in good part on mosquito larvae. Without the alligator they, along with other fishes and amphibians, would die off, but there would still be puddles enough to accommodate the mosquitoes. Multiplied by the million, they would plague the Floridians, not least the developers among them. More pesticides would probably then be introduced, but that would in turn take a toll of the birds, relieving swarms of other insects to intensify their ravaging of the state's crops.

Punch a hole in the ecological web, the moral is, and although the web may somehow knit together again, its life will not be the same—and neither, the chances are, will yours.

Besides the often vital service of an individual species and its present or potential contribution to man's physical well-being, the animals of the wild perform other functions that should appear useful even to the most self-oriented of humans. For one, they often act as monitors, warning the trained observer of danger.

In the tradition of the canaries that once signaled to miners a dangerous concentration of coal gas, certain mollusks are used to monitor the content of the water they live in. Individual bivalves are sampled to determine the chemicals present, especially mercury, in the water filtered by their food-intake process. The thinning of the brown pelican's eggshell, with a consequent decline in successful hatching, has similarly served to warn society of the dangers of DDT. The fragility and high visibility of butterflies have made them, likewise, extremely useful as monitors, their fluctuating numbers serving to indicate environmental change, especially in levels of pollution.

Still in the realm of the unproved, possibly even of the unscientific, is the belief that unusual animal behavior precedes an earthquake. Yet authenticated reports of strange animal conduct have been numerous enough for scientists to think them worth study—and even a hypothesis. Dr. Helmut Tributsch, a biochemist of the famous Max Planck Institute in Munich, investigated the actions of animals just before the earthquake in Friuli, Italy, in 1976. On that occasion, it was reported, deer came down from the mountains, crowded together near the village, and showed no interest in grazing. Cats, on the contrary, deserted the village en masse, with not one left in town when the quake struck. Cattle panicked in their barns, dogs started barking with no apparent reason twenty minutes before the quake, and chickens left their roosts.*

*Zoo animals have shown the same erratic behavior just before earthquakes. In San Francisco keepers reported that animals that normally mixed freely—zebras, ostriches, deer, and antelopes—separated and clustered in groups, species by species, minutes before the quake of August 7, 1979.

The explanation, Dr. Tributsch suggested in the esteemed British scientific journal *Nature*, is that animals may be sensitive to electrically charged particles produced by ground currents that precede an earthquake. The theory seems less far-fetched when we recall that certain types of weather, usually just preceding a storm and marked by high concentrations of positive ions, produce headaches and irritable behavior in some humans. The suggestion is still hypothetical, of course, but the editors of *Nature* regard it as "the most likely explanation so far." Whether or not it is true, animals can and do serve as earthquake forecasters in this way—though probably not enough people, contemptuous as humans are of any suggestion of animal superiority, have heeded their warnings to make an appreciable difference in the toll of lives.

From among the skeptical, exclude the Chinese, whose scientists have so healthy a respect for the predictive powers of animals that they systematically collect reports on unusual animal behavior. When these reached formidable proportions shortly before the earthquake of February 4, 1975, the government ordered tens of thousands of citizens to move from what appeared to be the focus of possible activity. The quake, when it came, was extremely severe, but the casualties were minimal. The saving in human lives was estimated in the hundreds of thousands. It should be added that such prophecies are not infallible—or at least not always timely enough. But they have been, at various times and in various countries, reliable enough for a Soviet geophysicist to call animals "the most sensitive disaster barometers known to science."

For breeders of livestock, the preservation of wild animals offers the prospect of new blood and new breeds in a sphere that is suffering, like agriculture, from a growing dependence on fewer and fewer subspecies. For these breeders the flora and fauna of the wild represent a gene bank of priceless potential value. The Natural Resources Defense Council suggests the possibility, for example, of upgrading Asian cattle by crossing domestic breeds with the selatang or Malayan gaur, an endangered wild ox of Southeast Asia. The Council also cites the success of the "beefalo" here at home. This cross between the bison and domestic cattle is said to produce a considerably cheaper beef, requiring less feed other than grass and reaching a desired weight in half the time.

The report of a task force sponsored by the Rockefeller Brothers Fund in 1977 summed up the case succinctly: "Each species (flora and fauna alike) is a potential resource of great value, since each one is its own unique biochemical factory. Such unique attributes can instantly appreciate in the scale of perceived human values from useless to priceless. Penicillium fungus is but one striking example." But it is the law of probabilities, the extrapolation of benefits already derived, that clinches the case.

So little is yet known about the plant and animal life of the world that experts' estimates gyrate wildly even as to how many species actually exist. On the *known* species of animal life there is of course substantial, though not perfect, agreement. The estimates of the Swedish zoologist Dr. Kai Curry–Lindahl, an expert for several pertinent United Nations agencies, are undoubtedly among the most reliable. He counts 4,200 species of mammals, 8,600 of birds, some 6,000 species of reptiles, and about 2,900 of amphibians. Other authorities number fish species at 23,000. Insects dominate all other forms of life numerically, with somewhere between 700,000 and 800,000 species.

What is less well known is that hundreds of thousands of additional animal species (with few mammals, if any, among them, but hundreds of thousands of insects) are believed to exist in remote areas of the earth and its seas, species that have yet to come under the microscope or even the naked eye of man. Every month a hitherto unknown bird or two is discovered by ornithologists beating the bushes of a South American jungle; a previously unknown form of marine life, such as a worm nine feet long, by the divers of a deep-sea expedition. The most informed estimate is that at least four times as many unknown animal species exist as those we know about, and of those we do know of, most still remain to be studied in any thoroughgoing way. Systematic study of the wild, it must be remembered, has a history of less than fifty years and must still be limited by the very need to avoid disrupting the habitats of the creatures under study.

The great objective, then, must be to keep species from dying out before exploration and study can even be fairly begun. Can it be counted as anything but insane recklessness to throw away whole species of organisms, any of which might well be our salvation in time of need, without knowing anything about them or their potential uses? Or to wipe out whole habitats, for that matter, without any idea of the life they contain? Who can say how many tragedies might have been added to this year's toll of human suffering if we had not learned of the medical possibilities in snakes and snails, in monkeys and molds? How much would have been lost to the development of genetics without the fruit fly, or to the science of evolution if the remarkable fauna of the Galápagos had been wiped out before H. M. S. *Beagle* brought young Darwin to those shores?

To borrow a phrase from the ecologist Erik Eckholm, there is "a presumption of value for every species, however obscure." He was talking, as we have been here, of tangible, material value. The vast intangibles, even greater in my view, are reserved for later consideration. For the moment it should be enough to think of Senator James Buckley's comparison between the obliterating of species and the burning of books. Allowing the species to disappear, he observed, is worse because "it involves books yet to be deciphered and yet to be read."

Extinct and About to Be

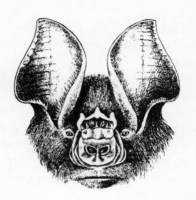

There were some little warm-blooded animals around which had been stealing and eating the eggs of the Dinosaurs, and they were gradually learning to steal other things, too. Civilization was just around the corner.

—WILL CUPPY

Man cannot be exclusively blamed for the vanishing of the world's wild-life; he has not always had a chance to contribute to the process. Ever since the beginnings of life on this planet, species have come into being, lived their allotted eons, and died out. No one can say even with the aid of computers how many animal species must have existed on the earth, but paleontologists, zoologists, and other scientists have come to a consensus that mammals, to take a single class, reached their peak of variety back in the Miocene, some 25 million years ago, and have been declining in number of species ever since. Man, to do him justice, has been around for only four million years or so, leaving nature to account for what happened in the previous 21 million.

Of all the mammalian orders* existing in the Miocene, 44 per cent

---

*An animal order is a large grouping that includes whole families, which in turn include a number of anatomically related genera. Each genus, in turn, includes various species. Man, to take the scale in a reverse direction, belongs to the species *Homo sapiens*, of the genus *Homo*. Other species of his genus were *Homo erectus*, *Homo habilis*, and *Homo sapiens neanderthalensis*. His genus belongs to the family Hominidae and the family to the order of primates, which embraces scores of species of apes, monkeys, and lemurs.

have disappeared. The paleontologist Pierce Brodkorb estimates that of all the bird species that ever lived—somewhere between a half million and a million—95 per cent are now extinct.

Why, then, the concern, the indignation about the prospect of further extinctions? What has always been presumably always will be, and there seems little sense in adding to the burden of guilt that oppresses at least some representatives of *Homo sapiens*. Alas, that is too easy on our species. When we look a little closer, it is apparent that it has had a great deal to do with the present state of matters zoological, a state that differs from all preceding ones not in the fact of change but in the startling *rate* of change.

In the age of reptiles—a time some 70 million years ago—dinosaur species, typically, were disappearing at a rate estimated by the ecologist Norman Myers at one every thousand years. The rate for all animals slowed down, or remained about the same, until the coming of man. Over the millennia of his existence—long by our sense of time but geologically the flick of an eyelid—he successively introduced fire, invented effective weapons, domesticated some animals at an ecological cost to others, and most devastating of all, appropriated whole habitats for farming, grazing, draining, and building. In consequence, all other forms of life have been disappearing at a steadily accelerating pace.

From the beginning of the Christian era, a recent event in the light of the earth's history, until 1650 about twenty species—ten of birds and ten of mammals—went into premature oblivion. That is, a species of one or the other of those two classes of animal vanished every eighty-two years. Since then, according to figures compiled by the biologists Lorus J. Milne and Margery Milne, the rate of extinction has been going up relentlessly and dramatically. Between 1650 and 1850, they calculate, an extinction occurred once every five years. In the following half century it was one every nine and a half months. Between 1900 and 1950, losses rose to one species every eight months. And these figures are for mammals and birds only. Add fishes, amphibians, arthropods, and insects—and why not?—and the toll has been incalculably greater. We have Dr. Myers' grim analysis that, counting all forms of life, plant as well as animal, we are now losing one species *per day* and that, with a wave of extinctions expected in the next decade because of further human inroads on habitat, we could by the end of the century be losing some species, from tree to insect, every *hour* of the day.

At that rate, how many animals would be left a century from now is not calculable, since so many thousands of species exist without our knowledge. But unless heroic measures are taken, the number is likely to be extremely small and to feature the kind of wildlife that thrives on man's leftovers: the coyote, the rat, the raccoon, the sparrow, the cockroach. Not quite the stuff of poetry and legend.

Statistics can be terrifying enough, but after a time they are merely numbing, like the numerical toll after a flood or earthquake. It is only when one thinks of individuals—in this case of individual species—that one appreciates the magnitude of the loss. When we learn that in the past two thousand years man has by his activities exterminated three per cent of all known bird and mammal species, half of them since 1900, we are left with a few sobering facts but little sense of the color of what has been lost.

To dwell for a bit on that color, the last passenger pigeon in the world died in September 1914, in the Cincinnati Zoo. Those who saw this end product of a race, bearing the commonplace name of Martha, may have thought it rather attractive, with its neck plumage running to iridescent greens, purples, and bronze. Or they may have thought it merely pathetic, living out its years as the last of its kind. It is unlikely that more than a few thought of the drama that had gone on before Martha's arrival in Cincinnati.

Less than a century before, the skies and woods of a continent were so dense with the migratory visits of her forebears that Audubon estimated their numbers not in the millions but in the billions:

> ... the air was literally filled with pigeons; they darkened the sun as in an eclipse, and their droppings fell like snowflakes. The whistling of their beating wings could practically make one fall asleep. ... During [their] evolutions the dense mass which they form presents a beautiful spectacle, as it changes its direction, turning from a glistening sheet of azure, as the backs of the birds come simultaneously into view, to a suddenly presented rich deep purple.

No one could conceive their swift disappearance from the face of the earth. But for food, for fun, and for the protection of farms they were clubbed in their nests, shot by the countless millions, eaten to the limits of human appetite, fed to the hogs or left where they fell for the foxes, raccoons, eagles, and vultures. Still greater slaughters occurred after the middle of the century, when railroads opened populous Eastern markets for the edible birds, and the last one left in the wild was killed in 1900. A race that had existed through eons of time had been dispatched by another species of animal in a matter of decades.

Just as we know when the last passenger pigeon took off for eternity, we know when two Icelandic fishermen extinguished the last great auk. That distinguished-looking bird, much resembling a penguin though not related, was known to be approaching the brink of extinction in the first half of the nineteenth century, primarily because of overhunting by sailors. Collectors and museums, more interested in relics than in living creatures, paid seamen to seek out the survivors on their last refuge—the island of Eldey, off Iceland. For several years they systematically raided the island for the flightless and easily taken birds, killing about a hundred of them to send back for stuffing.

Carefully bred in captivity, that number might have preserved the great auk for eventual reintroduction under an umbrella of legal protection. But spurred by a Reykjavik agent's offer of a hundred crowns per bird, a boat crew in June 1844 took the last two birds to be found on the rocky island.

The skins taken in those last years of the great auk showed up in museums around the world, and as time went on their value rose from the hundred crowns apiece that the sailors were paid to $20,000 for the relic of a vanished being. Could he have foreseen the future, the explorer Jacques Cartier would have been awed by the price when, only a few centuries earlier, he described how at a single landfall his men had killed more than a thousand great auks for their flesh and oil and still left enough of them ashore to have filled forty rowboats.

Not too different was the fate of the similarly flightless and fightless dodo on the island of Mauritius. Having had no predators to fear, the bird had long since lost its power of escape and so fell an easy victim to the clubs of the early settlers and sailors as well as to the domestic animals they had introduced, especially the pig. It was not particularly silly, as its name implied (*doudo* from the Portuguese for foolish) but merely helpless and tasty. The combination spelled its doom, as it did that of the giant moa, a spectacular wingless creature whose extinction, probably in the eighteenth century, likewise left the world less colorful. An ostrichlike bird that reached a height of from ten to twelve feet, the moa is still the subject of pleasantly chilling tales about "sightings" in remote spots of New Zealand, much like sightings of the Abominable Snowman in Tibet. The difference is that the giant moa very definitely existed almost down to the time of Captain Cook's visit and, legend has it, took a toll of Maori hunters with its lethal feet.*

The roll call of creatures recently gone into oblivion is by no means confined to birds, flightless or otherwise. It includes Steller's sea cow, a giant Arctic counterpart of the manatee, as long as thirty feet and as heavy as four tons; the quagga, a zebralike animal striped only on its forequarters, slaughtered by the Boers for food; and the thylacine or Tasmanian wolf. It is not certain that even now the thylacine has totally disappeared from the heavily forested reaches of Tasmania, but there have been no authenticated reports of a sighting in the past thirty years, despite rumors to that effect appearing from time to time in the Hobart newspapers. Whether or not a few of these largest of the pouched carnivores still exist, the island's leading zoologist writes that the thylacine "has reached the point of no return, and the most lavish attention will not save him."

Dr. Bernhard Grzimek, formerly director of the Frankfurt Zoo, makes the point that millions of dollars have been put up by the United Nations and private philanthropists to save the monuments of Abu Simbel from submersion in Lake Aswan, whereas "the preservation of the Tasmanian wolf, one of the most exciting and rare animals on the face of this earth, would require only a fraction of all this money." Like the giant sea cow, the quagga, and the moa, the thylacine is a far older monument than Abu Simbel, but the saving funds have not been forthcoming.

The disappearance of these creatures—I have not mentioned the Caribbean monk seal, Barbary lion, sea mink, Carolina parakeet, and several species of giant tortoise among the many that have passed out of the picture in modern times—may be numerically insignificant beside the wholesale extinctions that occurred in the early days of man's existence. But however many woolly mammoths and saber-toothed tigers man's remote forebears managed to kill, their departure, as species, owed more to the advance and retreat of glacial ice than it did to the pits, snares, rocks, and clubs of the early human hunter, even allowing for the thousands of huge creatures he may have killed by stampeding them over cliffs.

In any event those who would justify the behavior of twentieth-

*A comparable ostrichlike creature, the aepyornis or elephant bird, extinct since the seventeenth century, laid an egg equivalent in fluid to 180 hens' eggs. This Madagascar species was almost surely the source of stories of the fabulous roc, familiar to readers of the *Arabian Nights*.

century man by the behavior of his Neanderthal predecessors may have
a point, but it is not a point that on consideration they can want to press.
Nevertheless, the advances in technology, the geometric leaps in population, and the progressive obliterating of forest, jungle, plain, and marsh
—not to mention the poisoning of rivers and the very seas themselves
—all have combined to magnify so incalculably the threat to life in the
wild that the salvation of much that remains has become a question of
now or never. In the light of what we can expect in the very near future,
the speeded-up extinctions of the past few decades are a trickle foretell-
ing a cataract.

Two worldwide agencies now keep a running score on animal species clearly in danger of disappearing. In the town of Gland, Switzerland,
the International Union for the Conservation of Nature and Natural
Resources, acting through its Survival Service Commission, puts out the
closest thing to an official monitor of the world's wildlife—a series of
looseleaf volumes, to allow for changes, with red pages reserved for
creatures whose fate appears to hang in the balance. At present the *Red
Data Book*, as it is called, gives red pages (endangered) to 147 mammals,
80 birds, 54 reptiles, and 8 amphibians for a total of 289 animal species
(not counting fishes, insects, arthropods, and still lower orders). Their
"numbers have been reduced to a critically low level or the extent of
[their] habitat has been so drastically reduced that they are deemed to
be in immediate danger of extinction."

Even more stringent, surprisingly, are the appendixes to the international treaty known as CITES (for Convention on International
Trade in Endangered Species) adopted in Washington in 1973: surprisingly because these listings, going beyond the conclusions of the
concerned scientists who compiled the Red Book, are the law, in so
far as it is enforced, of fifty-one signatory nations. Appendix I to the
treaty lists those species so clearly threatened with extinction that all
commercial trade in specimens or their parts is prohibited except for
the purposes of propagating the species in captivity. Appendix II con-
sists of those animals—and plants—in sufficient trouble to require
the strict regulation and monitoring of trade so that a proposed export will represent no further threat to the species. Appendix III lists
species which a signatory declares to be endangered within its own
boundaries.

Putting aside cold statistics, let us look at the status of just a few
of the best known of these direly threatened animals, some of which are
not likely as species to survive the younger readers of this book.

ALL THE PRIMATES

Of the entire order of primates, every member species is on one or another of the convention's appendixes except for man himself, who so far as he can be sure is not near extinction. Which is to say, all his closest kin are endangered, threatened, or at the very least vulnerable.

In the worst plight are the great apes—the gorilla, chimpanzee, and orangutan. Indeed, among the most endangered of all large mammals is the mountain gorilla, the larger of the two varieties that make up the species. In spite of the genuine efforts of the Republic of Zaire to save these shy and inoffensive animals, their population now is well below a thousand and confined to a small mountainous area in eastern Zaire and Rwanda.

Adding poignancy to the gorilla's plight is the fact that, thanks to such students of its ways as George Schaller and Dian Fossey, we are just coming to learn how far it is from the maniacal chest-thumping demon it was for so long thought to be. Natives who from time to time kill female gorillas for food know that once the male troop leader has been trapped or killed, his followers yield, hands over their heads, to the clubs of their pursuers without even attempting to defend themselves. So consistent is this pattern that among the tribal hunters it is considered a disgrace to have been injured by a gorilla.

"The social interactions between members of a gorilla group," writes Dr. Schaller, "are close and affectionate, much like that of a human family, and their mating system is polygamous, a type for which man certainly has a predilection." Unfortunately, the resemblance is limited, humans being closer by blood chemistry, neurology, and behavior to the high-strung, often aggressive chimpanzee than to Dr. Schaller's "gentle giant."

If the chimpanzee is not in quite as much trouble as the gorilla, it is in trouble enough to be officially classed as vulnerable. The chief threat to the animal, next to the destruction of its habitat, has been the brutal and wasteful way in which hunters for years went about capturing baby chimps (as they did other primates) by first killing the mother. Since the average female gives birth to only three or four young in a lifetime, the drain has been severe, although less so than with the gorilla or the third of the great apes, the orangutan.

As for the "man of the forest," to translate his Malay name, the orangutan is disappearing as fast as the forests of his native Sumatra and Borneo. Barbara Harrisson, an authority on this primate, writes that "one thousand years ago there were far more great apes than men on Borneo" but now the estimated 2,500 to 4,000 orangutans on the Indonesian islands "are outnumbered by approximately three million people."*

Out of close to two hundred primate species in the world, nearly thirty are reported by *Focus*, a journal of the World Wildlife Fund, to be in danger of extinction. Besides the great apes already mentioned, the worst off of the primates are the lion-tailed macaques, down to five hundred or so in India's Western Ghats; the golden tamarin of Brazil, with no more than 250; some of the woolly monkeys; and all the lemurs, which probably most closely resemble our earliest primate ancestors.

Just as remoteness and local variations made the Galápagos Islands an invaluable laboratory for Darwin, so Madagascar, isolated and a thousand miles long, has developed a highly speciated fauna, especially among the prosimian lemurs. In that vanishing family are still such exotic creatures as the indris, the aye-aye, and the sifaka, rarely heard of outside of crossword puzzles. With the human species proliferating and the forest down to something between seven and ten per cent of what it was before it was opened to European settlers a few centuries ago, all these evolutionary links with the pre-anthropoid past, invaluable to anthropologists, are near to the vanishing point.

THE GREAT CATS

Anyone with the slightest interest in wildlife knows that the tiger, one of the most perfect weapons systems in nature, is the great cat with the dimmest prospects for a future in the wild. But it takes a benchmark or two to convey the speed with which it is plunging toward oblivion.

In the mid-nineteenth century some 100,000 tigers, divided into eight races or subspecies, roamed throughout Asia, from the Caucasus Mountains in the west to the mouth of the Amur River in eastern Siberia, and from Manchuria in the north down to the tip of India and the islands of Indonesia. One maharajah boasted as recently as 1965 of having all by himself accounted for 1,157 specimens in his career as a sportsman. There were many such sporting princes and in the days of the *raj*, many British aristocrats, too, who were happy to measure their mettle by the number of tigers they could shoot in a day from the vantage point of an elephant's back—comparatively safe in any case and particularly so with

Grzimek's Animal Life Encyclopedia, Vol. 10, p. 519.

hundreds of expendable beaters, guides, and professional hunters to intervene in emergencies.

Today fewer than five thousand tigers of all species remain in Asia, which is, of course, the only continent where they are to be found at all. In India and Bangladesh some two thousand of the Bengal subspecies remain out of the forty thousand that were there in 1900. Jungles from Malaysia to Vietnam harbor at most two thousand Indochinese tigers. Of the Siberian, largest and most impressive of the tiger races, only four hundred or so are thought to survive in Siberia, China, and North Korea, and only a few score Caspian tigers are left, mostly in Iran. Of the Indonesian races, a few hundred Sumatran tigers remain, and the Javan subspecies is down to five or six specimens as this is written—which is to say that, like the Bali tiger, it is already gone.

Such chance as *Panthera tigris* has for survival in the wild it owes to a genuinely, if belatedly, concerned Indian government (Prime Minister Gandhi's early authoritarianism was at least a boon to wildlife) and to the World Wildlife Fund. The tiger was the beneficiary of that organization's first concerted and major effort to save an endangered species. Encouraged by the Fund's pledge of a million dollars and abashed by the wanton slaughter of animals following the departure of the British, the Indian government designated nine existing wildlife sanctuaries as tiger reserves, putting up some $4.5 million to improve them and to begin the dismantling of villages that had grown up within park limits. The move was described by the project's director, K. S. Sankhala, as "a nucleus for regenerating the species, a sort of series of nurseries for tigers, where they can rest, breed, spread out."

The Fund's contribution went to increase park personnel and supply them with equipment to fight poachers: jeeps, pickup trucks, motorcycles, and even such specifically adapted pursuit vehicles as saddled camels in the desert areas of Rajasthan and jet speedboats for the delta areas of the Sunderbans. In addition, research scientists were supplied with dart guns and radio telemetry equipment to track animal movements.

Though all this effort may prove fruitless in the long run because of the constant pressure for land, it has served so far to stabilize the Bengal tiger population—in Nepal and Bhutan as well as in India. There is at least a chance that the tiger will not join the long list of animals that once roamed India but have vanished from that subcontinent—in ancient times the hippopotamus, the great apes, the giraffe, and very recently the cheetah. In 1951 the last three cheetahs in India were killed

off, though they were once so populous there that a sixteenth-century
Mogul emperor kept a thousand of them in the royal stables to use in
hunting, an activity for which they are readily trained.

From a habitat that once ranged across most of two vast continents,
the cheetah is now almost exclusively an African animal, shrinking fast
and classified as vulnerable even in its last stronghold. A survey made
by Dr. Norman Myers when he was regional officer for Africa under the
United Nations Food and Agricultural Organization indicated roughly
ten thousand to twenty-five thousand still on the continent, with a good
chance that the figure will be halved in the next few years and the game
reserves unable to protect more than three thousand of them. The much
more endangered Asian species is now confined to a handful in southern
U.S.S.R., Iran, and Afghanistan.*

Even a casual traveler in East Africa must sense that the leopard,
though increasingly hard pressed in specific areas of its broad habitat,
is in nowhere near the dire plight of the tiger, a much less adaptable
animal. After several weeks of hard riding by jeep and by elephant in
the wildlife reserves of India, I was extremely lucky to see one tiger in
the wild (strictly speaking, one-third of a tiger through the crisscrossed
branches of a bamboo thicket); by contrast, in the first of a few days
spent in Kenya and Tanzania I was easily able to get snapshots of
leopards at their ease on the branches of thorn trees. I saw one loping
across a road not a hundred feet ahead of the vehicle I was riding, and
another unhurriedly ambling over the lawn of a lodge where I was
having breakfast. Indeed, the animals are so close to Nairobi that in
Karen, a suburb on the border of Nairobi Park, I was told by my hostess
that she could not leave her two huge dogs outside the house at night
to guard the property. Leopards, she said, get out of the free-roaming
park from time to time—evidently no difficult feat—and help themselves
to a meal of dog meat, of which they are especially fond.

The snow leopard, a species distinct from the true leopard—an inability to roar is one difference—is very close to extinction. Theoretically
protected in India, China, the Soviet Union, and Nepal, it is in grave
danger all the same, largely because its luxuriant fur will bring a hand-

---

*A census of cheetahs is considerably harder to take than one of tigers, which stay put
in their own territory, whereas the cheetah roams far and wide in pursuit of herbivores
that follow the forage. Within a restricted area, moreover, tigers can be kept track of by
the pugmarks they leave, which are as different for each animal as thumbprints are among
humans. The cheetah, unlike other cats, does not have retractile claws and so leaves a track
complicated by clawmarks.

some price to a poacher capable of setting traps, snares, or poisoned stakes along the animal's regularly used trails. The best estimate is that the population of this exceptionally beautiful creature is down to a meager five hundred.

WOLVES, BEARS, AND PANDAS

To leave for the moment those Felidae who are about to leave us, other endangered carnivores fabled in most of man's cultures are almost all varieties of the wolf, and several major subspecies of the bear.

As a symbol of wickedness in the lore of the Western world, the wolf owes its reputation far more to its one-time ubiquity and its occasional forays on domestic livestock than to any consistent record of savagery toward humans.*

At one time wolves were common throughout Europe and Asia, from England to Japan, and in the western hemisphere from the Arctic to Mexico. Today there are sizable populations only in Siberia, Canada, and Alaska, with perhaps another thousand left in Minnesota. Elsewhere in the United States the timber wolf is gone from 99 per cent of its former range. The red wolf, which once flourished throughout the southeastern part of the country and as far west as Texas and Louisiana, is, if not extinct in the wild, probably the most endangered of United States mammals. Besides the usual pressure from farmers, hunters, and builders, the red wolf seems to have been a peculiar victim of assimilation. As its population dwindled, it took to mating with coyotes and feral dogs, with the result that after several generations very little of the genetic quality of the animal remains.

Apart from the wolf's major ecological "use" in keeping its prey population—mostly deer and rodents—under control, it holds out promise of rewarding research on at least two fronts significant to man. Its highly sophisticated social organization, especially as related to pack hunting, is thought by some scholars to offer better clues to the behavior of early man than that of the apes. Just as remarkable, if true, is the

---

*It is true that although there is no record of people being killed by wolves in North America, many such casualties occurred in Europe and Asia, sometimes from rabid wolves. In a region of central France, more than a hundred authenticated attacks, most of them fatal, occurred in a three-year period of the eighteenth century; but that unexplained wave seems to have been unique. Wolves followed Napoleon's army in its retreat from Russia, and in a similarly grisly way the animal's population increased again in Western Europe, according to *Grzimek's Animal Life Encyclopedia,* following the ravages of World War II.

animal's possible population-control mechanisms, which are believed by some to keep its numbers in balance with the space and carrying capacity of a changing environment.*

What the lion was to the Africans and the wolf to the American Indians, the bear has been to the Eskimos and Aleuts of the Far North —that is, the ultimate test of a hunter's prowess, the subject of song and story, the symbol of power. Of all the world's bears, the most awesome and impressive are the grizzly, the Kodiak, and the polar—and all three are in trouble.

The Kodiak, actually a subspecies of the brown bear, is four feet high at the shoulder and nine or ten feet when it rears up on its hind legs. For eons this largest of the world's carnivorous land mammals has enjoyed the relatively mild climate of a few southern Alaskan islands and a strip of the mainland, but it has increasingly been made to share those lands with a growing cattle enterprise. Caught between militant cattlemen and trophy hunters, the bears have found sanctuary only in the Kodiak Wild Refuge, which has been a blessing to it but not quite enough of a blessing. The density of bear population in the preserve is extreme now and is expected to intensify under the Native Land Claims Settlement Act.

Both the grizzly and the polar bear are listed in the appendixes of the international convention on endangered species. Of an estimated one hundred thousand grizzlies in the United States (not counting Alaska) in the early days of the republic, there are considerably fewer than a thousand left, most of them in Glacier and Yellowstone national parks. The polar bear, thanks to its forbidding habitat, is somewhat better off, but until a few years ago hunters with snowmobiles, airplanes, telescopic rifles, and a willingness to pay a guide $2,000 for the thrill of killing one in perfect safety were continuing to reduce its numbers seriously.

To an extent as yet undetermined, the polar bear is now protected by an agreement, effective in 1976, among the five polar countries: Canada, Denmark (for Greenland), Norway, the Soviet Union, and the United States. Killing polar bears is at least officially outlawed except by native peoples for their own limited use—and in that event by traditional methods only. The hope is that the population of these animals will be stabilized at its present level, estimated by C. Harrington, a close

---

*I have found this claim as vehemently rejected by reputable zoologists as it is defended. It is mentioned here largely as a subject for profitable further study.

student of the polar bear, to be somewhere between 12,000 and 20,000. Any number below that minimum is likely to doom a creature whose widely dispersed habitat would make mating too uncertain to sustain the species.

Until a comparatively short time ago, the giant panda was generally believed to be a species of bear, although some thought it closer to the raccoon. Taxonomists now classify it as neither, tentatively assigning it a genus all to itself while allowing the probability that it is related to the species known as the lesser panda.

For an animal that had never been seen by Western eyes until 1913, the giant panda has been quick to endear itself. Because of the all but unapproachable nature of its habitat in the thick and high forests of China's interior, nobody can say how many of these seemingly cuddly animals (though actually it would be foolhardy to try cuddling a creature with such powerful teeth and jaws) may exist, but authorities tend to put the number at something over a thousand. The unusually severe winter of 1976 wiped out much of the bamboo on which giant pandas feed, reportedly costing the lives of many of them. In any event the animal is rare and would most likely be doomed if it were not for a promising joint effort to save it by the Chinese government and the World Wildlife Fund under the direction of Dr. George Schaller, on loan from the New York Zoological Society.

ELEPHANTS AND RHINOS

The African elephant's plight is a special case, full of sad ironies. Some wildlife preservationists believe it is not really threatened at all, but according to Iain Douglas–Hamilton, a foremost authority on the subject, we are in the final phase of elephant history. And if Dr. Bernhard Grzimek, author of *Serengeti Shall Not Die,* is right, that phase will not be long: "Elephants have to disappear whether it is in ten or fifty years." There are at least a million elephants in Africa, yet the *Red Data Book* lists them as vulnerable nonetheless, and they are so listed on Appendix II of the convention. Finally, and more ironic than these seeming contradictions, is the fact that although the African elephant's only hope is the national parks, these same parks have now proved to be its undoing.

Two circumstances make for the elephant's dim prospects. The simple and short-term factor is the price that can be had for its tusks. The wholesale price of ivory went up from about $2.80 a pound in 1969 to $48 a pound in the mid-1970s, primarily because of the unstable monetary conditions prevailing in the world. At the lower price the demand had been met largely from "found ivory"—that is, from the tusks of animals that had died a natural death or had been killed by park wardens to reduce population pressures in a given area. But once the commodity was regarded as a hedge against inflation, Kenya and Tanzania became a poachers' gold mine.

Unfortunately, the official protection of elephants in Kenya was exceeded by protection of the poachers. The country's elephant population was cut in half between 1970 and 1977, a disaster perhaps inevitable since the wife and daughter of President Kenyatta had a major investment in the illegal traffic and profited handsomely. Under the succeeding administration—and following a modest drop in ivory prices—poaching decreased somewhat, but it is still a lively business in spite of the treaty forbidding the importing of ivory by signatory countries.

Much ivory from the African elephant's great incisors until recently found its way to Hong Kong and other points, whence it emerged eventually as necklaces, bracelets, chessmen, and *objets d'art*. Some of it reaches the world's markets by way of Alaska, masquerading as finished products from walrus tusks.

Grave as are the pressures from poaching, they are less so than those of a shrinking habitat. The great parks of East Africa have become havens for threatened elephants retreating before the poachers, farmers, and grazers of the region. Their range, says Douglas–Hamilton, "is now confined to islands of wilderness lapped by seas of humanity."* It is not a situation that can last.

The African elephant's somewhat smaller and more readily trained Asian cousin is even more immediately threatened. Here, too, the problem of a shrinking habitat is compounded. The Asian elephant has for centuries been the workhorse, tractor, and bulldozer of India and Southeast Asia. When its numbers were great, it was more economical to draw on the wild for fresh labor than to breed elephants already in service. For the most part females are used, the males being unreliable, even dangerous, in their recurring periods of *musth*, a glandular condition related to breeding that may make an animal uncontrollable. But a pregnant female is more or less unusable for eight or ten months of the approximately two years of gestation. During that time, of course, she

---

*The slow strangling of animal populations in well-intended reserves that are necessarily too small will be more fully discussed in the next chapter.

eats like the elephant she is, requiring a considerable input for a long-deferred return.

Even more of a financial drain is represented by the offspring, which must be fed and cared for over a span of ten years or so before it is sufficiently developed and trained for a career as a work elephant. Accordingly, centuries of depleting the wild elephant population plus the proliferation of farms, towns, and tea plantations at the expense of elephant habitat have decimated the animal's numbers, leaving little likelihood of the species' survival beyond the next few decades.

Unless drastic rescue operations succeed, the next largest of the land mammals, the rhinoceros, will not outlast the century. Only the so-called white rhino—until recently headed for disaster but now strictly protected by the South African government—is given much of a chance, with a present population of about three thousand.

The black rhinoceros, not long ago to be found everywhere in Africa, has become the most swiftly vanishing of all the earth's great mammals, some 90 per cent of East Africa's specimens having been slaughtered by poachers in a ten-year period. Some idea of the rate of killings may be had from the observation in 1979 of a ranger in Tanzania's Amboseli Park. Of the seven hundred black rhinos in that preserve in 1975, he told a reporter, "there are five very scared animals left, trying to survive."

Of the three Asian varieties, including the great one-horned Indian rhinoceros, there are all told not more than 1,200 left, a thousand of them confined to the sanctuaries of Chitawan in Nepal and Kaziranga in Assam. No one who has seen these gigantic armor-plated beasts gently grazing the reeds along the Brahmaputra River can believe that the world ought willingly to exchange such living relics of the Miocene for stuffed exhibits in a museum of natural history. But if the Asian species is to be spared this fate for a time, the pressures for land and the genetic hazards of inbreeding with so small a population would seem to put severe limits to the length of the reprieve.

Even so, it should outlast the African black rhinoceros, which has to contend not only with habitat problems but, even more, with ancient superstition and human greed. Because of its phallic shape, the horn has for centuries been absurdly valued in the Far East as an aphrodisiac and a cure for fever. For just as long, rhinoceros horns were often made into cups thought to be capable of warning the drinker of the presence of poison. By the fall of 1979 the horn of the poor doomed rhino was bringing well over $300 a pound wholesale (a horn might weigh two or three pounds) to be sold for much more as ornate dagger handles in the markets of the Middle East. In the single year 1976–77, the New York *Times* reported, the lives of 4,000 rhinos were required to meet the *macho* demands of Yemen's dandies. The price of the African horn is modest enough compared with that of the Asian, which commands $1,000 a pound in Thailand and almost any figure you can name in the black markets of Hong Kong.

LOST AT SEA

We have concentrated for purposes of illustration on the status of large animals—the great apes, big cats, huge bears, elephants, and rhinoceroses—because as a general rule the larger the animal, the more vulnerable it is to extinction. Because of their size they make the best targets, require the most extensive habitats in a world increasingly preempted by humans, and are the slowest to reproduce. Larger than any of the creatures we have mentioned is the blue whale, and not surprisingly it is the most threatened of all.

If the run of human beings had ever seen a whale, they would not use words like *elephantine* and *mammoth* to convey a notion of size, for a blue whale is larger than thirty elephants. Its heart alone weighs as much as six husky men, and the whole of it weighs four times as much as the largest dinosaur that ever walked the prehistoric earth. A creature of such magnificent proportions ought to need no argument for continued existence, but it does all the same.

To their infinite harm, the great blue and its kindred species as well —the humpback, gray, bowhead, right, fin, sei, sperm, and others—have provided man with more than second- and third-hand excitement. They have been living sources of raw materials for the making of soaps and scents, lubricating oils and gelatin, corset stays and umbrella ribs, and hair oil. They provide leather for handbags and shoes; fertilizers; food for pets, livestock, and humans; and a score of other products, from glue and candles to lard and linoleum.

The terrible consequence is that, in spite of alternative sources for all those products, one of these great creatures is still being killed, with official sanction, every half hour of the day and night—about 16,000 a year under quotas fixed in 1979 by the International Whaling Commission. Since a number of whaling countries do not belong to the Commission, the total number killed is at least twice that. The United Nations Environment Programme has put the estimate for surviving blue whales at fewer than a thousand. Other authorities think there are two or three times that number; but even if they are right, the prospects are that too few whales will find each other across the oceans of the world to keep the rate of births above the rate of deaths.

The whole cetacean family, including the little dolphin, offers the very real possibility of some day revealing to man a means of vocal communication other than his own, and all the whales are of a high order of intelligence. But whether or not the humpback's singing is even potentially a language, as some think it is, we may never know. Indeed, we may never know anything of the large-brained whales beyond their dimensions and their store of products. "The moot point," wrote Herman Melville, "is whether Leviathan can long endure so wide a chase, and so remorseless a havoc; whether he must not at last be exterminated from the waters, and the last whale, like the last man, smoke his last pipe, and then himself evaporate in the final puff."

The animals mentioned so far in this chapter, glamorous as they are, represent only a small part of the color that will in all likelihood be lost to the world within the next few decades. Others just as familiar and just as doomed are the jaguar, puma, Asian lion, crocodile, and gaur, to cite at random. But the excitement of the animal world is by no means confined to creatures of such renown. For the more knowing, there will be as great a loss in the vanishing of scores of animals less well known but every bit as fascinating.

Vulnerable but not quite nearing extinction are the ocelot and many other smaller cats of the wild, the giant armadillo, several species of otter, and those dubious imitation mermaids, the dugong and the manatee. Except in zoos, the wild horse is gone and the wild ass is not expected to be with us long. Neither is the tapir, the okapi, or the wild Bactrian camel. Many species of deer, sheep, goat, and antelope are nearing the end of the road, including the bighorn sheep, the markhor, the oryx, the addax, and other such creatures already more common in Scrabble games than in life.

The birds-in-trouble list is similarly long and sad. Familiar on the roster of the doomed are the California condor, the Philippine eagle, the Eskimo curlew, several woodpecker species, and some of the most beautiful of the cranes—the whooping and the Siberian. Dozens of parrot and parakeet species are on the danger list, all of them bright and colorful, and so are varieties of petrels, eagles, pheasants, pigeons, and warblers.

Among the reptiles, all the sea turtles are endangered, especially the green, hawksbill, olive ridley, and leatherback. American alligators, under the mantle of protection, have made something of a comeback in Louisiana and Florida, but the crocodilians in general—from the Nile to the Orinoco—appear to have come to the end of a road that has taken them millions of years to travel.

So much for the present state of life in the wild.* How did it come to reach that state and why do the wilderness, great parks, and reserves offer less hope for the future than animal-lovers can wish?

*For an extensive list of creatures approaching the brink, readers are referred to sections of the IUCN's *Red Data Book* reprinted at the end of this volume.

What Wild?

Wherever Man has spread his dominion, scarcely any flight can save or any retreat harbour; wherever he comes, terror seems to follow, and all society ceases among the inferior tenants of the plain . . .
— OLIVER GOLDSMITH

Anyone with a feel for wild animals must want them to remain, first and foremost, in the wild. Not because their natural habitats constitute a tranquil, secure, and plentiful haven for them—some sort of Eden where they live out their allotted life spans in relaxed comfort. Quite the contrary. Forever trying either to make food of other animals or to avoid being made food themselves, they lead lives which by sentimental human standards must seem harried and cruel. Generally those lives are much shorter than those of creatures in captivity, given their zoo counterparts' certainty of daily food, freedom from predators, and medical programs ranging from obstetrics to dental care. No, it is the very hardships of life in the wild that in a remorseless Darwinian way make wild creatures what they are, what nature intended them to be, and what man instinctively wants them to remain. As individuals they may do better in captivity, but as species they do better in their own ecosystems —*provided there is enough of those ecosystems left to sustain them.*

Tragically, it is too late in the day to ponder such abstract choices. We are on the last lap of a race between those who with guns and traps are removing animals from the wild and those who with bulldozer and

axe are removing the wild from the animals. In a rising order of importance, the drains on wildlife are the bag of the individual hunter, whether
he kills for dinner or for sport; the take of the commercial hunter,
whether legally engaged or a poacher; and, most significant by far, the
encroachments of a relentlessly growing and land-hungry human population.

The noncommercial hunter rates no more than passing attention
here, because extensive food-hunting is confined to smaller and smaller
numbers in shrinking areas of the world, and because sports hunting has
almost everywhere come under sharp limitations and controls. In fact
the noncommercial hunter performs, to some extent, an offsetting service to wildlife.

At the lowliest level the pursuit of game by remnant hunting-and-
fishing societies for meat on the table is here and there a factor in the
decline of wildlife, but by no means a principal one. The Pygmy hunter
of central Africa, using traditional weapons, far from being a threat to
the animal life of his region, is himself an integral part of its ecology;
his hunting is an essential contribution to its natural balance. The same
is true of the Eskimo, for example, when he hunts the seal or the caribou,
though less true when, equipped with snowmobile and explosive har-
poon, he kills more specimens of the fast-vanishing bowhead whale than
he can consume. Elsewhere the drain is greater, but it is impossible to
fault protein-hungry people for saving themselves from malnutrition; it
is nonsense to condemn them on humanitarian grounds when the rest of
the world sees nothing immoral in killing cows for their beef, lambs for
their chops, and geese for their livers.

Killing for sport, at least in these days, does not pose a very much
more serious menace to the world's wildlife than shooting for food,
though obviously it is a great deal less innocent. Where species were
seriously depleted by noncommercial hunters in times past—wolves,
pumas, and bears come readily to mind in this country—the ostensible
purpose was to rid the burgeoning settlements of "varmints," as the
cowboy films have it, rather than to provide outdoorsmen with a pastime.
But to concede the point is not to endorse the wanton and needless
slaughter of the "varmints," and certainly not the killing of animals for
the fun of the thing.

Such hunting reached scandalous proportions in Africa, India, and
Southeast Asia in the heyday of imperialism, and in the United States
of the nineteenth century as well. At India's Bharatpur bird sanctuary

a billboard lists some of the bags taken when the haven was still a private duckshoot for the maharajahs of that once princely state. As late as 1943 a party that included Viceroy Sir Archibald Wavell knocked down 2,310 birds in a single day. Yet it is fair to say that the worst slaughter of wildlife in the country's history came in the months following independence. Freed from controls as well as colonialism, the Indians went berserk, shooting game wherever they could spot it, with guns often distributed by the new government to encourage crop protection at a time when food was scarce.

In time successive Indian governments, along with those of other recently liberated countries, took a very different tack. Yet even as he announced a complete government ban on big-game hunting, Kenya's Minister for Tourism and Wildlife conceded in 1977 that the previous year's legal bag had accounted for only six thousand animals, compared with the vastly higher take of the poachers. The Minister's counterpart in Tanzania went further. At a 1979 press seminar in Costa Rica on the international trade in endangered species, the Honorable Solomon Ole Saibull acknowledged his country's debt to the sports hunters of the past, most of them from Europe and America: "They hunted for pleasure and they wanted the enjoyment of that pleasure to last. So it was in their interest to assist us in protecting the animals, and they did so as they hunted."

For centuries this same principle of conservation-for-pleasure provided a rationale for the hunting nobility of Europe. They made a point of saving land and limiting the take so that they might continue indefinitely to enjoy the glories of the chase, even if that meant hanging peasants who poached a little meat for the pot. It was pointed out to me by C. G. C. Rawlins, director of the London Zoo, that thanks to the combination of this aristocratic self-interest, subsequent game laws, and the needlessness now of hunting for food, there are more deer in England today than there were in Robin Hood's time.

Richard G. van Gelder, of New York's American Museum of Natural History, goes so far as to say that "there would be no big game left had it not been for hunting and the royal game preserves." It is likewise estimated that there are more white-tailed deer in the United States today than when Columbus reached the New World. Similarly, the wild turkey's numbers are up considerably from the turn of the century, and so is the elk population.

Among the conservationist and environmental organizations that do not oppose strictly regulated sports hunting, though they may have little enthusiasm for it, are the Sierra Club, the Defenders of Wildlife, and the National Audubon Society, whose founder, George Grinnell, was an avid hunter—as, indeed, was the great Audubon himself. Aside from what they have done in the way of preserving habitats and promoting regulation of the take, self-interested hunters argue that, like other predators, they perform the function of keeping the herds of prey animals healthy by thinning out their ranks. As serious a conservationist as Thomas L. Kimball, executive vice-president of the National Wildlife Federation, has stated flatly in his organization's journal that "there is no realistic way for deer populations in Wisconsin or Wyoming to achieve a self-sustaining level without hunting unless wolves, coyotes, and mountain lions are restored to their former abundance" which, he needlessly adds, "is not likely to happen."

To these contentions of the hunters there are challenges. Just as automatic rifles are a far cry from the muzzle-loading shotgun, not to mention the blunderbuss, so jeeps, Land Rovers, snowmobiles, and powerboats allow the sportsman to penetrate far more quickly and deeply into the haunts of his prey than was remotely possible in the days when he depended on a horse, sled, elephant, or his own two feet to get him to the scene. With sites far more penetrable, and with weapons now so overpowering, the odds against the hunter's prey have grown proportionally.

What is worse, the thrill of the sport is to take not the weakest, the smallest, the most sickly of the herd, as a carnivore might, but the biggest, the most glamorous specimen, the one whose head will look most spectacular on the wall or whose hide will most nearly cover the floor. Hear the young Theodore Roosevelt, whose home in Oyster Bay is still a nightmare assembly of heads and horns:

No sportsman can ever feel much keener pleasure and self-satisfaction than when, after a successful stalk and good shot he walks up to a grand elk lying dead in the cool shade of the great evergreens, and looks at the massive and yet finely moulded form, and at the mighty antlers which are to serve in the future as the trophy and proof of his successful skill . . . There have been few days of my hunting life that were so full of unalloyed happiness as were those spent on the Bighorn range . . . always amid the most grand and beautiful scenery; and always after as noble and lordly game as is to be found in the Western world.

The danger in this selective hunting is that even though a species survives, it may do so in a gradually more stunted form, the "fittest" in this case being those with the least appeal to the vainglory of the hunter.

It appears remarkably easy to persuade oneself that the most glamorous prey shares the pleasures of the chase with its most Rooseveltian predator. The aptly named J. A. Hunter, in his likewise aptly named book, *Hunter*, concluded regarding photographic safaris:

> *The day may come when the camera will take the place of the gun in African hunting. In many ways, it will no doubt be a fine thing. Yet I am glad that I lived in a time when a man went out against the great animals with a rifle in his hands instead of a device to take pictures. Sometimes I think the animals themselves liked it better, too.*

Other hunters take a less mystic stance, however, viewing their prey not as glorious creatures at all but rather as mean brutes whom it is their duty to eliminate. In a hunter's guide to Indochina, published in Saigon only a half century ago, I came across this quaintly indignant observation on the tiger: "Being a masterpiece of dastardliness, it will seldom give the hunter a chance of a clear shot, unless on a bait"—which, the guide notes elsewhere, is necessary "to meet the unsportiveness of those cowardly pests."*

More common, and in the long run probably more damaging than this arrogant ignorance, is the sheer mindlessness of the "slob hunter" who shoots any moving target, injures animals without bothering to track them down and finish them off, and freely distributes his trash throughout the wooded area of his operations. Nor can it be overlooked that if hunters prevent an overpopulous deer herd from starving by substituting themselves for the carnivores that once kept that population in check, it was human hunters who upset the balance in the first place by exterminating the carnivores. Which is to say that if hunters save some species, they do so too often at the expense of others.

With respect to deer, moreover, credit to huntsmen for the preservation of species can easily be overdone. A major contributor to their increase is the leveling of high-ceilinged primary forests, where ground forage is far scantier than it is in the low, secondary forests that succeed them. The thinning out of the woodland has promoted the health of deer

*From *Notes on Big Game in Indochina*, by G. Tiran, distributed by the Anglo–American Tourist Bureau, Saigon, 1929.

populations—though not always of the forests themselves—as much as the hunter's services in keeping those populations in balance for his own pleasure.

Even one who takes a dim view of sports hunting, however, must concede that, compared with commercial hunting, the harm done is minimal. In spite of the warnings of conservationists and the considerable legislation already on the books, international traffic in wildlife reached catastrophic proportions in the 1970s. In 1978, five years *after* representatives of eighty nations had met in Washington to head off imminent disaster from that traffic, the United States alone imported $12 million worth of reptile hides, $69 million worth of feathers and down, $140 million in furs, and close to $8 million in ivory. In addition, three million live mammals and reptiles and some 600,000 live birds were imported— not more than 1.2 per cent of them destined for the country's zoos. Overwhelmingly the catch went to pet shops, with a comparatively modest take awarded to research.

Positive results of the Washington convention (CITES), only now beginning to appear, offer a fair degree of hope. But evasions, legal and illegal, are numerous and the volume of trade beyond the established control machinery remains appalling. A single dealer in Thailand can still expect to handle over a million birds a year, counting both foreign and domestic sales. As recently as 1976, Indonesia alone was still annually pouring into the world markets 28,000 crocodile skins, 270,000 monitor lizard skins, 350,000 snakeskins, and 71,000 tortoiseshells. With the coming of rapid air transport, the devastating losses en route, particularly of birds, declined sharply, but for that very reason importers found it more profitable to engage in the trade. Overwhelmingly such live birds —mostly parrots, finches, mynahs, toucans, and hummingbirds—are absorbed by the pet trade: that a parrot costs $25 in Mexico and sells for $200 in the U. S. amounts to an open invitation to do business. Only rarely are these captive birds bred. The result of the traffic has been a great drain on avian populations, sometimes an outright loss of species, in such exporting countries as India, Thailand, Pakistan, Singapore, and China.

The obstacles to control are formidable in spite of energetic and worldwide efforts to keep threatened species from being wiped out by commercial exploitation of jungle and sea, savanna and marshland. Most of the world's wildlife, obviously, is in the developing countries, which are otherwise the poorest, lacking the funds to support well-armed, well-paid, well-trained men to enforce their game laws and patrol their reserves. The modern poacher, backed by adequate funds from abroad, conducts major operations with trucks, jeeps, machine guns, light aircraft, and, not least, the wherewithal for bribery. Not long ago sixteen illegal hunters were picked up in a raid in Tanzania, and all sixteen, it turned out, were game wardens. Not much wonder when a single reptile skin will yield a park official as much as six months' work and a rhinoceros horn possibly as much as two years'—without the risk of being murdered for overzealousness in line of duty. It is in the light of this lopsided battle that one can appreciate the World Wildlife Fund's efforts to bring aid and equipment to the forces of the law.

On balance, little of the blame for the illicit animal traffic rests with the countries of origin. As long as big sums of money are held out for the skins, horns, hair, and other parts of wild animals, ways will be found to get around the law and around the excellent efforts of private organizations and governments to control the traffic. Only half the world's countries are now parties to CITES, which has actually been in force only since July 1975. Those on the outside—notably Belgium, Mexico, and Thailand—often act as conduits for animals originating elsewhere.

Forged permits and other papers complicate the lives of customs officers in the signatory countries, handicapped as they are to begin with by a layman's inability to identify parts of animals as in fact coming from animals on the proscribed lists.

Reptile skins and cat skins are particularly hard to distinguish; so, to a lesser extent, are birds and bird products. Taking advantage of the prevailing ignorance, a zoo official in Kobe, Japan, a few years ago went so far as to instruct a potential seller in Indonesia to shorten the tails of some birds of paradise, color them here and there with water paints, and ship them out as magpies.

It is, in Russell Train's words, "a dirty and sordid business," but prevailing prices guarantee a traffic, whatever means are required. We have seen that a pair of moderately heavy tusks may bring a dealer thousands of dollars (of which the poaching group will get probably a third). An American lynx skin, which brought no more than $40 in 1973, was quoted at $340 six years later, with those of the more luxurious Russian variety priced as high as $1,800.* Hyacinth macaws smuggled into the United States are known to have brought $5,000 and more

---

*Because strains in U.S.–Soviet relations recently kept American buyers from the fur auctions in Leningrad, a full coat of Russian lynx cost, in 1980, anywhere from $100,000 to $150,000.

apiece, and Arab sheiks addicted to falconry will pay seven or eight times that much for a rare gyrfalcon.

Conservationists have called the sea turtle the most profitable animal left in the wild. Tortoiseshell from the hawksbill species, used for jewelry, rivals ivory in value. Parts of the inoffensive creature's skin make excellent shoes, and the meat and cartilage too often wind up in gourmet soups. The eggs not only are a food item but are believed in Latin America to be an aphrodisiac, always bad news for an animal species.* The result is that creatures that would allow a Spanish galleon to navigate in a fog just by the sound given off by their huge migrating herds are near extinction, and most of their huge rookeries no longer exist.

While large smuggled consignments of endangered animal products are beginning to come under control, thanks to CITES and its monitoring and information agency, known as TRAFFIC (Trade Records Analysis of Flora and Fauna in Commerce), a more difficult source of wildlife smuggling comes from the swelling streams of tourists. Innocently, or at least ignorantly, they get through customs gates by the thousands with souvenirs that not long before were parts of living animals.

In a move to discourage this aspect of wildlife exploitation, the government of Kenya in 1977, having banned all hunting earlier that year, went on to prohibit the sale of animal souvenirs. A deadline was fixed and scores of shops in Nairobi were compelled to clear their shelves of horns, hides, tusks, zebra-tail fly swatters, and similar relics. If governments around the world followed suit, customs inspectors would no longer pass, knowingly or unknowingly, such doubtful necessities of life as whale-tooth scrimshaw, hornbill snuff bottles, tortoiseshell guitar picks, and tiger-tooth good-luck charms. Somehow a tourist who knows that a vicuña coat comes from vicuñas seems unable to make the connection between an elephant-foot wastebasket and a real elephant's foot.

As a threat to the world's wildlife, rampant commercial hunting competes, of course, with other hazards—notably the same chemical pollution of the earth's air and water that is beginning to undermine the health of man himself and the heedless use of pesticides that destroy the food chain, from insects through birds to mammals. But hazardous

---

*Besides turtle eggs and the much-publicized rhinoceros horn, alleged aphrodisiacs include the antlers of the sika deer, crocodile kidneys, ground spiders, the glands of the musk deer, and tiger steaks. But there is no evidence that all of these taken together will fulfill expectations—the ingredients appear to be rather less potent than the myth.

above all is the incredibly swift disappearance of room on the planet for wild creatures to find their food, pursue their prey, flee their enemies, breed, and otherwise assure the perpetuation of their respective species. It is this fast-shrinking living space that changes the pertinent query from "Are animals best left in the wild?" to the starker question, "What wild?"

In the saving of animal species it has become apparent that conservationists cannot deal with them one at a time. Except for emergency action to prevent the immediate dying out of, say, tigers, gorillas, and whooping cranes, they cannot profitably set out to assure the future for individual species—because far more than individual species are involved. It is whole ecosystems that have to be saved.

"You can't deal with three million species on a one-to-one basis," says the ecologist Thomas Lovejoy. "You have to protect them in the aggregate, and the only way to do that is to protect entire biotopes." These regions of uniform environment have to be large enough, if individual species are to survive, to sustain populations capable of avoiding inbreeding. Some studies of the subject conclude that it takes roughly a thousand breeding animals to assure the required genetic diversity of a species, but there is no real agreement.

But what is happening to these habitats? In the past decade the world's population has gone up by about 746 million people, while all classifications of its wild and uninhabited acreage, as we shall see, have decreased spectacularly. The correlation is clear. More people need more land on which to live, travel, raise crops, graze domestic animals, and do business. They also need more of the resources to be found in the wild —timber from the forests, minerals from the ground, water from the lakes and rivers. All this comes under the head of progress, no doubt benefiting man in the short run, but in the only slightly longer run injuring him gravely by dooming countless species of plants and animals on which he might one day have to depend.

Going beyond biotopes, it is whole biomes that are under attack, a biome being one of the several major types of the earth's area, each marked by a characteristic climate and vegetation—such as tundra, savanna, and desert. And of these the hardest hit is the tropical rain forest.

With very small fluctuations in moisture and temperature and, until recent decades, a minimum of human intrusion, the rain forests of the world have spawned by far the greatest variety of plant and animal species to be found on the planet. Through millennia they have been its major gene pool, from which most species have emerged to colonize the world. With only a fraction of that potential remaining they have already gone into a swift, probably disastrous, decline in area and diversity of life. The National Academy of Sciences estimates that across the world, tropical forests are being cleared at the rate of 50 acres for every minute of the day and night. That is 72,000 acres a day, *27 million* acres a year —or 42,187 square miles. Which is to say, an area of rain forest bigger than the entire state of Pennsylvania is disappearing annually.

So far, two-thirds of Asian forest land has lost its original tree cover. South America, according to estimates of the Food and Agriculture Organization of the U. N., retains only 36 per cent of what was once its moist forest. Madagascar forests are down to 6 per cent of what they once were. In the thirty-five years since independence India has cleared something like a quarter of its jungle area, and Indonesian forests have become virtual quarries for teak loggers. And the terrible irony is that very little of the land taken from the forests has turned out to be good for farming: the soil degenerates quickly once the growth–decay cycle of the forest is broken, its nutrients leached out by the rains.

In Brazil, which all by itself has been losing an annual slice of forest land the size of Indiana, that lesson is just now being learned, with the possibility that in the Amazon basin, some of the world's last giant forests may yet be saved. Two huge national parks and a biological reserve have been created, covering an area of some six million acres— though that is still small compared with the damage that has already been done. "Probably the most crucial of all conservation problems today—aside from the continued growth of sheer human numbers—is the accelerating destruction of the wet tropical forests of the world, particularly in the Amazon basin," says Russell E. Train. "At present rates of destruction these forests will be largely gone within this generation, and we will have presided over the elimination of a very major proportion of all the species of life that have evolved on the face of the earth since the Creation."

Other major habitats that have suffered severe modification and outright loss in modern times are the prairie and the wetlands. Half the Western prairie of the United States has gone down before the plough of the farmer and the herds of the livestock growers, following into oblivion the Eastern tall-grass prairie, which now covers less than one per cent of the area it once did. Wetlands around the world have fared just as badly, with marshes, bogs, swamps, and mudflats generally considered wasteland, to be drained and turned into resorts, industrial developments, marinas, shore drives, and the like. California has lost an estimated half of its original wetlands and the Mississippi Delta about a quarter. It is the world's wetlands, of course, that provide temporary haven and feeding ground for millions of waterfowl and other migratory birds, apart from performing such other vital functions as recharging aquifers, absorbing potential flood waters, and affording breeding ground for fish and amphibians. The drastic reduction of wetlands has caused a severe loss in the world's stocks of fishes, amphibians, rodents, reptiles, and mollusks of all sorts, not to mention the herons, egrets, and other birds that feed on them.

As great and distinctive areas of the earth—forest and savanna, woodland and prairie—shrink before the advances of man, their animal life diminishes along with them. In the past few decades the process became perceptible enough to induce governments, however belatedly, to set aside acreage for national parks, bird sanctuaries, and natural reserves—a patch here and a patch there to keep species of plants and animals from extinction, to save for the people of an increasingly paved-over civilization a link with the awesome world out of which they emerged.

As one lucky enough to have visited a dozen or so of the major animal reserves of Africa and India, as well as most of the national parks of the United States and a handful in Latin America, I found very few of them unrewarding and many exciting beyond all expectation. I would hate to have missed, in particular, the thrill of my first sight of Ngorongoro Crater, that surviving slice of the world of Genesis seen through the rising mists of morning. Nor am I likely ever to forget the solid pink walls of flamingos at Kenya's Lake Nakuru, accurately described by Roger Tory Peterson as "the greatest ornithological spectacle on earth"; or the swift currents of wildebeest, zebra, and antelope streaming across the Serengeti. I would not willingly have missed the prehistoric one-horn rhinoceroses browsing along the Brahmaputra, best viewed from the back of an elephant in Kaziranga Animal Reserve near the border of India and Bhutan. Nor the tree-climbing lions of Lake Manyara, the Cahuita Rain Forest on the shores of Costa Rica, nor, among the most vivid of all such recollections, the painted storks perched like fixed wooden figures atop the scrub trees in the marshes of the Bharatpur bird sanctuary in India.

Yet for all the glories of these animal havens, the pleasure they afford is accompanied by rising doubts as to their adequacy, their durability, and their capacity to serve as the sole custodians of the world's disappearing wildlife. No such doubts can detract from a conviction of their enormous value, the need to expand them greatly and to keep them going as long as possible in a world that is filling up all too rapidly with one species of animal at the expense of all the others. But the realities encourage no grandiose hopes for them.

Consider first the area involved. After decades of grave concern and extremely hard-won victories by conservationists, roughly one per cent of the earth's 55 million square miles of land is devoted to what the United Nations lists as National Parks and Equivalent Reserves. And of the entire protected area, a forbidding percentage is in the remote and species-poor regions of Greenland and northern Canada.

This is not to belittle the relatively impressive acreage allotted for the purpose by a small but growing number of states with significant wildlife still on hand. Ten per cent of Tanzania is in national parks, including the magnificent Serengeti, the Ngorongoro, and the newer Selous Wildlife Reserve, which alone covers 21,867 square miles—close to the area of Massachusetts, New Hampshire, and Connecticut combined. Kenya can take almost as much pride as Tanzania in this respect, and similar credit goes to other African states whose concern is the more commendable for their devastating poverty. In tiny Israel, hard pressed as it is for living space, some four per cent of the land is invested in protected areas, including the imaginative Hai Bar reserve, which is in the process of reassembling as many as possible of the 130 animal species mentioned in the Bible.

Yet, tragically, all the effort and money that have gone into habitat preservation, from the Arctic to the edge of the Antarctic, have not been nearly enough to cope with the need, or to spread the attention effectively around the globe, or even to keep the protected areas themselves from deteriorating. The sad truth is that many of them are parks and reservations in name more than in fact, for everywhere they are steadily the worse for the intrusions of surrounding populations, the competition of domestic animals, and erosion by tourists—without whose interest and money, admittedly, there would probably be no park systems at all.

It is the surrounding populations, with their hunger for farm and grazing land, that seal off the migration routes of the Serengeti elephants, forcing them back on the "protected" territory. There they have little choice but to destroy vegetation which would otherwise have time to regenerate in natural cycles, to starve, or to break out and wreck the huts of poor farmers in the borderland. Those that remain in the park reduce vegetation to a point where it is no longer adequate for their own needs or for the needs of other animal species. This excessive destruction of trees, moreover, allows a runoff of water that produces the kind of dried-up land that already characterizes a good part of Kenya's Tsavo Park. In short, instead of serving as expansion belts for periodic migrations, the settled bands bordering the great African reserves have become cordons slowly constricting the life of the parks.

More directly, man and his works simply invade the world's animal reserves, either to graze their cattle or to settle in as squatters bent on permanent colonizing. In the middle of India's Sariska Game Sanctuary, for example, I was dumbfounded to witness a noisy religious ceremony attended by a throng of worshipers at a shrine in a large clearing, where no grass grew but litter was abundant. It was as though a revival meeting were being held in the midst of Yellowstone. But nothing could be done, I was told, because the shrine was there first. More startling, Erik Eckholm quotes in a Worldwatch paper the report of a South American analyst that "there are 30,000 *campesino* families living within the national parks, forest reserves, wildlife refuges and protected zones" of Venezuela. "This illegal squatting then becomes *de facto* agricultural land and is subsequently legalized."

Foraging for wood in the Indian parks is even more common than squatting. The country's reserves suffer from a demand for timber that grows almost by geometric progression, and ironically, the very saving of forest land is part of its endless cycle of problems. The less timber available for fuel, the more that need must be met by cow dung. But the more cow dung is used as fuel, the less there is left for fertilizing the fields and the greater the volume of chemical fertilizer that has to be imported at prohibitive cost and used at considerable risk. In some areas, fast-growing eucalyptus trees are being cultivated to break the cycle, but for the foreseeable future India's poor, when not raiding the forests, will go on burning the fuel that is to be seen stacked in carefully shaped and patted mounds along every road, its combustion giving the country's atmosphere its characteristic pungency.

In other, less direct ways, the activities of man on the margin of the parks bear negatively on the chances of wildlife survival. A case in point is Costa Rica, a beautiful country which, in spite of the poverty that afflicts so much of the tropical world, has made heroic but often futile efforts to protect its natural heritage. It is hard for at least this traveler to forget his approach to the Reventazon River. Scrambling down a path from the height of Turrialba through one of the few remaining primary forests, I was thrilled with the beauty of the fast-flowing river between the dark heavily wooded hills—until I got close enough to see that what I had taken for white water was in fact the white of scummy wastes, carried by a stream that is made to serve the region in place of prohibitively expensive treatment plants. Worse than the esthetics of the thing, the river empties its content into the Caribbean not far from the tropical rain forest of Cahuita, where the tides eventually bring the muck ashore, along with local garbage, to the considerable detriment of that otherwise fascinating feature of the land. The river is heavy with pesticides, too, as are other Costa Rican streams, whose chemicals, carried offshore, have noticeably darkened the colors of the coral reefs.

At least one of Costa Rica's reserves, Santa Rosa, is plagued by a human activity that threatens parkland throughout much of the underdeveloped world—the grazing of man's domestic livestock. A constant battle is waged on the cattle-growers—complete with fines, jail sentences, and sometimes guns. But it is hard to carry on a war against cows when the export of beef for the fast-food counters of the United States is a major factor in the country's economy. The more hamburgers consumed by North Americans, the less protected woodland for Costa Ricans and their now far from plentiful fauna. Ocelot, jaguar, peccary, and tapir are still to be found, not to mention a hundred more bird species than are to be seen in all the hemisphere north of Mexico, but the strain grows harder as the population mounts, and prospects dictate more hope than assurance.

What is true of Santa Rosa is true to a far greater extent elsewhere in the world. Domestic cattle are far harder on the earth than their fellow creatures in nature. Wild ungulates do not duplicate one another's food tastes on the Serengeti plain: one herbivore grazes the low grass, another the high; some browse the foliage of shrubs and short trees, others the tall—with additional variations to allow for differing bodily requirements. Thus the available vegetation is enough to sustain a wide assortment of animals. By contrast, the tastes of domestic animals tend to overlap. According to a study by the IUCN, "The result of the nonduplicating food preferences of the mixed wild ungulates is that virtually all of the available vegetation can be used efficiently to support the biomass of mixed herbivores. Whereas, when domestic livestock graze, only one class of food—grass—and only a few species within that class, provide the bulk of the preferred forage and most efficient source of nutrition." As a result, the study continues, *equivalent savanna lands support a biomass of wild ungulates that is two to fifteen times higher than that of domestic livestock. The higher biomass from wild ungulates leaves the savanna in excellent condition, while the domestic livestock grazing in East Africa—even on managed ranches—virtually always result in some degree of overgrazing and consequent depletion of the land's productivity.*

Yet on the fringes of game parks everywhere, it seems, cattle are pressing on. Harold T. P. Hayes, author of that impressive book on East Africa, *The Last Place on Earth*, notes that in the world today, "there is a larger biomass of cattle than of men; in Asia, cattle consume more protein than men." Ngorongoro has long been victimized by the incursions of Masai tribesmen, for whom cattle are not only the source of blood-and-milk nutrition but in their numbers the very symbol of wealth and prestige. At one time as many as 13,000 Masai were allowed to settle within the bounds of the Ngorongoro Conservation Unit, though technically, at least, outside the crater itself. Their way of life is not likely to diminish the problem for conservation authorities, what with a population increase of close to three per cent a year.

The Masai are far from being the only tribe in East Africa with an envious eye on the pastoral opportunities of the great parks. From a Tanzanian National Park Report,* I cull a few items of relevant interest:

Goats and a herd of cattle were caught grazing in Arusha Park ... There was a problem with the Barbaigk tribe in Tarangire in that they refused to settle in planned villages and resorted to taking cover in the park ... four Barbaigk were caught grazing their cattle in the park ... In May a herd of nine cattle and forty-two goats were caught while grazing in Arusha and the person concerned was sent to court ... In Tarangire, one Barbaigk was caught grazing a herd of about 1,000 cattle at Chubi, within the park. ...

In India, of course, the sacredness of cows has added a major dimension to the concern of conservationists. There, bony cattle everywhere contribute a spiritual grace, an economic drain, and an ecological disaster. By their ubiquitous overgrazing they have, among other achieve-

*Excerpted in *Africana*, Vol. 5, no. 12, 1975.

ments, made inroads into the deer population in the tiger's habitats. Crowding out the chital, a spotted deer which was for long a staple of the tiger's diet, they at least share responsibility for the great cat's approach to extinction.

As though the world's national parks and reserves are not sufficiently threatened by those to whom they nominally belong, there are always visitors to compound the damage. By its very nature, tourism is a drain on the integrity of these pockets of wildlife, for all that they could scarcely survive for a year without the dollars, pounds, marks, and yen that the travelers bring in. For the more fortunate of the world's developing countries, tourism can make the difference between economic hardship and outright disaster.

The irony is that what is for a time a triple boon—to government, to tourists, and to wildlife—may be, and generally is, a boon of steadily diminishing returns. The naturalist Aldo Leopold long ago explained why very simply in *A Sand County Almanac:* "All conservation of wilderness is self-defeating, for to cherish we must see and fondle, and when enough have seen and fondled, there is no wilderness left to cherish."

That phenomenon is recognized around the world and accounts for the skill with which ambivalent "natives," from Cape Cod to the Cape of Good Hope, combine a low esteem for visitors with a high esteem for their cash. They know, too, the axiom of modern travel, that the more rewarding a tourist attraction is today, the less rewarding it will be next year and the year after that.

Cases in point are universal. In the African game parks even the seemingly innocent pursuit of snapshots has already altered to a considerable extent the life of the animals. Their food-hunting is frequently interrupted, a serious matter in view of the great number of attempts a lion or a cheetah has to make before it scores a kill. It is reasonable to wonder, moreover, whether the sex life of a pair of lions inhibited in the center of a circle of Land Rovers is better than that of lions in a well-ordered zoo.

Norman Myers tells of the disruption in the natural routine of crocodiles on the Upper Nile, where tourist launches deliberately race their engines to induce the frightened animals to leap into the water, sometimes abandoning their nesting ground to watchful baboons intent on making off with an egg or two. The resulting photos are exciting—but the crocodiles, which of course have survival problems other than tourism, are for the first time in their hundred-million-year history faced with the threat of early extinction.

Closer to home, our own national parks are under devastating pressure, with a quarter *billion* visits a year. Tourists wilder than anything else in the area nearly took over Yosemite a few years ago, introducing motorcycles, crime, and the less than natural sounds of acid rock. Other national parks were in a fair way to be overrun by private cars until the number of vehicles was gradually forced down, none too soon to prevent air pollution up to the very rim of the Grand Canyon.

The problem is how to allow more and more people to see the glories of the wild without having them overwhelm, transform, and in the end destroy the objects of their admiration. In the remote Galápagos Islands an effort is being made that those in other natural areas might do well to emulate. Tourist facilities there are extremely limited, and the traffic is held down in accordance with the fragility of each island. Some may be visited by no more than ninety visitors a day, some by no more than twelve, all under the guidance of carefully trained personnel. Some may be visited only by scientists with special permission and others are altogether barred to human intrusion. It should be noted, however, that whereas the Galápagos have to deal with 12,000 visitors a year—the number is controlled by the Ecuadorean authorities—640,000 visited Kenya's parks and reserves in 1974 and possibly 200,000 those of Tanzania.

India wants very much to increase tourism in its game parks—has to, indeed, if it is to go on supporting them against competitive demands on the land—but its attractions present some formidable difficulties for the casual tourist. Except for Sariska and the glorious Bharatpur bird sanctuary, which is hardly twenty miles from the Taj Mahal, the game parks are in remote corners of a huge country, reachable as a rule only by endless bus rides over roads often fascinating, but sometimes washed out and always jarring, dusty, and exhausting. Outside the cities, where comfortable hotels are generally available, lodging ranges downward from the barely tolerable.

More significantly, perhaps, India's animals, in particular the spectacular ones that most interest the tourist, are hard to see and harder still to photograph. Whereas those of East Africa are mostly creatures of the plain and can be viewed by the thousand, those of India are animals of the forest. One can ride a jeep or an elephant all day and all night through Corbett National Park in Uttar Pradesh, for example, and be rewarded with nothing more exciting than a distant shot of one more cluster of axis deer or perhaps, with luck, a "blue bull" (nilgai) in the distance.*

These very difficulties may benefit the few tigers left in Kanha, the rhinoceroses in Kaziranga, and the other wildlife glories of India—among them the sambar and barasingha, the huge gaur, and the brilliantly plumed jungle fowl, ancestor of our barnyard chicken. But the Indian government is surely right in assessing the need of its parks for tourist money, with all the risks involved. Without some financial return the parks will be too great a luxury to allow any government to turn a deaf ear to other claims on the land and its resources.

A conservation-minded political leader in what the South African government holds out as a "homeland" for its Zulu population was quoted a few years ago on the point in the IUCN Bulletin: "More and more of my people . . . see my enthusiasm for the wilderness getting less and less relevant to the major issue of their survival." Governments in Africa, Asia, and Latin America may perceive their game parks and reserves as sources of tourist income, but to the wretchedly poor they are the playgrounds of rich foreigners.

With all the television pictures of Africa's savannas and lush-looking forests, it is probably not widely appreciated that only ten per cent or so of the continent is suitable for farming, with all the better lands already under cultivation. The dubious marginal areas now occupied by wildlife are all that remains available, but even they look good to pinched tribesmen on their borders. Kenya's population goes up by 3.5 per cent a year, which means that the acreage of even moderately arable land will surely have disappeared by the end of the century. Enviously contrasting their scrawny cows with zebras grown sleek on the savanna grass of the parks, Kenyan tribesmen will sneak in a bit of pasturage wherever they can get it, legal or not. At the very least they are entitled to resent, even with passion, the raids of marauding elephants on their wretched farm plots, the destruction of a meager maize crop, and certainly the occasional attack on one of their family defending a field. In such cases they are all too ready to kill an in-

---

*Elephant-riding is the best way of seeing India's wildlife and, for excitement, a genuine advantage over the African Land Rover—but it can be carried too far. Riding an elephant astride, as I did in Kaziranga—in order to avoid being caught on the wrong side of a howdah for a possible camera shot at a tiger—can be harrowing. Cramps work their way up from one's widely extended feet, to the calf and thigh, finally communicating to the brain the distinct impression that one is doomed to go through life looking like a croquet wicket big enough to accommodate an elephant.

truder capable of eating five hundred pounds of food a day—and they are legally entitled to do so.

Or consider the population pressures in India, which in its thirty-odd years of independence has increased in absolute numbers more than the United States has added in all the two hundred years of its history. In spite of drastic efforts to reduce the birth rate, the country grows by the population equivalent of Australia every year—all seeking a little land of their own. In Southeast Asia, one can bewail the hacking down of forests by the lumber companies of the industrial nations, but to hungry Indonesians assured of even temporary work in building access roads and hauling logs, a job is likely to outweigh the loss in orangutans.

Whether or not the Indonesian government should allow the exploitation of its forests or the Costa Rican government should allow the sale of beef to compromise the integrity of its wildlife reserves, both have pressing need of foreign exchange and both have impoverished populations whose individual short-range needs will inevitably be served. It is this dilemma that more than any other single cause, makes the future of wildlife parks doubtful, not to say bleak.

Interviewing William G. Conway on the subject, I found him pessimistic, as most authorities are, about the prospects for these parks, but not altogether so. Distinctions must be made between those like Nairobi, in which automobiles seem to breed more rapidly than anything else, and Amboseli, which Conway's own New York Zoological Society is working hard to improve and support. Some reserves, he thought, would remain untouched for a long time because the nature of their terrain makes them undesirable for human uses—desert, high mountain, and similarly uninviting habitats. A high proportion of Africa's game park area, he suggested, is not of much long-range use for other purposes.

But these are glimmers of hope in a picture which Conway, like others, has from time to time described in gloomier terms. "The future of wild animals *in nature,*" he observed in 1971, "is generally poor. . . . The point of near extinction will probably be reached by most Asian faunal communities within five decades: in Africa and, especially, in South America it may take but little longer if present trends continue."

Other experts, thinking of population pressures and the development of drought-resistant strains of maize, which could be grown in some of the least fertile of the protected areas, see even dimmer prospects for Africa's reserves. Norman Myers, who believes the Serengeti even now ought to be more than doubled in acreage, points out that the contrary is more likely:

*By the end of the century, even with extensive family-planning
campaigns and with intensified agriculture through green revo-
lutions, there will be at least another 13 million people with no
land in their ancestral reserves and with no employment prospects in the urban conglomerations that will then pass for cities.
Therefore, these people will look for any patch of country on
which to plant subsistence crops—woe to wildlife if that is wildebeest, or zebra, or gazelle habitat.*

And, he adds, the same situation is true in Zambia, Ethiopia, Mozam-
bique, Angola, and Tanzania. Rhodesia, South Africa, and Namibia "are
already developed to the point where wildlife is reduced to token communities in isolated parks and reserves."

The most dismal forecast of all comes from a man who believes that
only by cropping and harvesting wild game can a balance be maintained
that can save the parks, save the animals (though many fewer of them),
and save the people by giving them a share in the "harvest." A complex
battle has long been raging between the proponents of Human Management and those of Leave it to Nature, with men of good will ranged on
both sides, but with all agreeing on one proposition—that human pressures drastically threaten the wildlife reserves, no matter what. It was
on this score that Ian Parker, a professional cropper and leading spokes-
man for the Human Management faction, told Harold Hayes:

*We are heading into an extended dry period. Assume the human
population doubles and triples there and the people somehow
manage to eke out survival until a really classic drought comes,
the sort to which East Africa is given. Their only hope will be to
graze the Serengeti. When that point comes, no government
elected by the people is going to deny those people access ... The
Serengeti will be eroded. The whole phenomenon has nothing to
do with wildlife. It has to do with human density.*

Harvesting—a repulsive term, implying that intelligent beings like
chimpanzees and dolphins are to be thought of in the same way as
stringbeans and alfalfa—rests on the economic concept that a species
will be saved if there is sufficient financial reward for saving it. The
"take" need only be limited to what is called the optimum sustainable
yield for a reciprocal benefit to accrue: a continuing revenue for the
people of the area from the sale of animals or their parts, and a continuing incentive to preserve the species as the very source of their liveli-

hood. Like any other predator, the reasoning is, humans will have an interest in keeping their prey in balance, to the perpetual advantage of its population and their own. Thus will animal species be preserved by man's enlightened self-interest.

In one degree or another a substantial number of conservationists take this view. Among the most articulate of them is Norman Myers, whose work with the Food and Agriculture Organization of the United Nations repeatedly brought him into contact with the hunger so tragically prevalent in Africa. In Kenya's Tsavo Park in the last few years, he reported in 1975, at least 7,000 elephants had died of starvation, meaning an incalculable quantity of meat "left to the vultures at a time when one quarter of a million people in Kenya received famine relief food." Elsewhere he was quoted on what nourishment might be spared from the hoofed herds of East Africa: "The Serengeti migration could produce perhaps 40 million tins of canned meat each year without any decline in the total wildebeest, zebra, and gazelle numbers below the present two million."

I recall hearing Solomon Ole Saibull, Tanzania's Minister for Natural Resources and a highly respected conservationist, hold forth to the same effect:

> *I think we have dwelt at sufficient length on the nasty aspects of international trade in wildlife and their products. I believe there is something positively good about the trade and it would be inadequate to ignore this other aspect. Wildlife and their products are a resource which can be used to improve the lot of people in those Third World countries which possess that resource. Wildlife products and proceeds can clothe people, provide them with food and shelter, either directly or through foreign exchange earnings from the sale of live creatures and their products.* *

It was a sentiment I was to hear many times in the course of my investigations. "Harvest your surplus," said Gerald Lentz, manager of the Busch Gardens. "If that's what it takes to save animals, then why not?" Others pointed out how cropping and culling had for decades increased the numbers and well-being of ducks and deer while creating economic markets in food and hides. And Lee Talbot, recently appointed director general of the IUCN, echoed the view in making clear the position

*From an address to the Earthscan Seminar, San José, Costa Rica, March 25, 1979.

of the World Wildlife Fund, when he was that organization's conservation director, in connection with a temporary need to cull vicuña. The organization, he said, was "not opposed to the utilization of wild species if this conforms to sound conservation policies, is based on solid science, and is humane."

Among conservationists who oppose harvesting, the fear is that "a legitimate trade opens the floodgates to illegitimate poaching." To which Roger Caras added the point that both the fur industry and all too often "the very highest seats of government" itself, are too much motivated by greed to be reliable partners in a legalized business supposedly based on sustained yield.

The experience of the whaling industry tends to reinforce such doubts, providing powerful evidence that in the animal trade, as in others, immediate maximum profits take precedence over sustained maximum yield. In a grim determination to amortize their investment in ships and equipment, whaling countries seem altogether heedless of how long the source of their income can continue. Despite quotas set by the International Whaling Commission, Japan and the Soviet Union, the two leaders in the industry, have so depleted the population of their quarry that the prospects for profit are disappearing as fast as the whales.

Whatever the merits of managed and harvested wildlife, the more of a fact it becomes, the less of a difference there must ultimately be between humanly controlled animal reserves in a so-called "wild" and the great zoos of the world. Except, as succeeding chapters should show, that the zoos will be able to provide, in addition, a wealth of scientific study, sound breeding practices, and, above all, an easy access to wild animals for great numbers of people—an access basic to a broad public understanding of the role of animals in the life of man.

Changing Zoos: The Need to Breed

There is reason to hope, should the great wildlife forests be lost, that the world may be grateful for even a few seeds. That is where captive propagation of wild animals should go from here.

—WILLIAM G. CONWAY

Some time ago the President of Burma offered to make London's zoo the gift of a takin, a rare creature resembling a muskox but thought by some taxonomists to be closer to the chamois. To the President's surprise, the gift was politely declined by the secretary of the London Zoological Society, later to become Lord Zuckerman. The secretary suggested, however, that since the animal could still be bred in Burma, in spite of its scarcity, he would be glad to accept a *pair* instead.

Nothing was to come of that counter-proposal, but much has been heard of the principle behind it—that a zoo should not serve as a showcase for rare animal specimens unless it intends to promote an increase in their population. Offered only a single specimen, Lord Zuckerman has said, he would turn down even a coelacanth.

Of the score of zoo directors and general curators I talked with across the United States and Europe, none supposed that, even following that principle, zoos could rescue and propagate all the earth's endangered species. Some, like Dr. Theodore H. Reed, director of the National Zoo in Washington, suggested that two hundred species might conceivably be saved, but most entertained more modest hopes, ranging down to

Dr. Conway's estimate of perhaps a hundred. What breeding programs require, above all, is space, and he points out that all the zoos of the world put together are currently obliged to make do with an area smaller than the borough of Brooklyn. Nevertheless, there is the same unanimity of opinion that zoos are obliged to do all they conceivably can —some say it has become their primary reason for existence—to save as many species as human effort and the budding state of breeding technology will allow.

Biologically, captive breeding may be like raising livestock, but in the present state of knowledge about the mating habits of wild animals it is not much beyond where livestock-raising must have been in Abraham's time, and thus filled with the color and excitement of surprise. After all, it is only in the thirty-five years since World War II that the great majority of zoos have stopped offering themselves simply as places of entertainment. Even the London Zoological Society, which had an early eye to science, made its name first as an attractive amusement. Indeed the new word zoo came into its own only when a music-hall artist of the 1860s tickled London with a song that went, "Walking in the Zoo is the O.K. thing to do."

As for the animals that were the show in most early zoos, they were cooped up in noisome little cages, and although they sometimes survived to a respectable age, it is doubtful whether they enjoyed their lives, if one accepts the reasonable view of animal behaviorists that satisfaction in animals is demonstrated not merely by longevity but also by breeding, appetite, and the absence of pathological behavior.

Anyone over the age of fifty is likely to recall the zoo trips of childhood as a montage of balloons, hot dogs, and rows of somewhat scroungy animals, either sleeping or eternally pacing back and forth in small, smelly, almost bare cages—except for the primates, which often took out their boredom in public masturbation or in aggressive efforts to shower the nearest bystander. Even the ungulates in their outdoor paddocks were lethargic, and only the seals could be counted on to offer a mild but continuing entertainment.

In the spring a few babies would arrive, each good for a feature story in the local papers showing mother lion and cubs, baby hippos destined soon to wallow in a foul indoor tank, or perhaps a fawn being bottle-fed by the smiling wife of a keeper. If the seasonal crop was insignificant, it hardly mattered, since animal dealers were always on hand to fill gaps in the collection with freshly imported specimens. All of this was a far cry from the beauty that characterizes the best of today's zoos, and a still further cry from their concentration on propagating rare species. Not all of the vast improvement in zoo-keeping that has marked the past forty years stems from the necessity of collections to renew their own animal populations, but I am convinced that this need to breed is the key to most of what has been happening.

By the mid-1940s many of the zoos from England to Japan had been wrecked or starved out. Frankfurt was down to twenty animals when the great Bernhard Grzimek took it over. With the job of restocking, zoo people everywhere became acutely aware of what the wisest of them already knew, that populations in the wild were drastically low, and were due to become less and less available to zoos as newly established states in Africa and Asia either slaughtered their wildlife without restriction or began to clamp down on its export. For the first time animal counts were undertaken from the air, along with field studies of breeding habits and animal behavior in the wild, many of them under the aegis of the world's great zoological societies. All of which produced a great ferment among zoo personnel everywhere, and a surge in breeding programs that in turn inspired advances in zoo building, animal grouping, veterinary medicine, ethology, and other scientific research. As a result, the great zoos of today are as far removed from the menageries of fifty years ago as computer technology is removed from the methods of the Victorian counting house.

Zoos now generally breed their charges with three specific ends in mind. The first and most obvious is to relieve pressure on the wild, as zoos move from their old role of wildlife consumer to that of producer —a purpose forced upon them in part by national and international regulations that make it increasingly hard to import wild animals. The second is to raise animals with a view to reintroducing them into the wild, a matter to be discussed later in this book. And the last is to serve as an ultimate haven, the scene of a last-resort effort to save and propagate endangered species as a scientific and aesthetic boon to future man. If museums can give viewers a sense of awe with plastic dinosaurs and lifeless mastodons, how far more thrilling for our grandchildren to see, in all their flashing vitality, the last of earth's grizzly bears, snow leopards, lemurs, and bongos.

SAFETY IN NUMBERS

A captive breeding program begins—but only begins—with numbers. J. M. Knowles, director of the private and remarkable Marwell Zoological

Park in Hampshire, is unambiguous: "In no case can it be satisfactory from a conservation viewpoint for a zoo to maintain merely a single pair, trio, or even a small representative species group. Any serious breeding effort must involve substantial numbers, either within the one collection or, preferably, among several cooperating institutions."

Within a particular zoo the case for numbers is obvious: let one of a breeding pair or even one of a small group die, and unless it is soon replaced—no longer an assured thing—the species, as far as that zoo is concerned, is doomed. But even with no such misfortune, the much broader consideration of inbreeding remains. The narrower the genetic base of an animal group—and the fewer infusions of new blood—the more likely that the genetic weaknesses of the original pair, if any, will be perpetuated and the adaptive variability of the species be lost.

This is too simply put, no doubt, and there is some room for debate among the experts on the subject of inbreeding. But in the main none of them disputes the general proposition that in zoo breeding programs safety lies in numbers. Perhaps the most dramatic, though hardly the most significant, illustration of the effects of inbreeding is the case of the white tigers, which have long been a major attraction of the National Zoo in Washington, although some zoologists there and elsewhere are inclined to be more apologetic than proud concerning this particular exhibit.

The story of the white tiger goes back to the time when the Maharajahs of Rewa, one of the princely states of India, enjoyed a degree of independence under the British raj. In Rewa, hunters would at long intervals bring in for their ruler a white tiger which was neither an albino nor a different species from the usual Bengal variety; it was simply a mutation, but a singularly beautiful one, with chocolate stripes on a white coat and blue eyes instead of the usual yellow-green. The skins made striking rugs and wall hangings.

After independence the princes fell on hard times, and the Maharajah of Rewa, who at first was allowed to keep his lands, was persuaded that the sale of live white tigers would be worth many more rupees than their skins alone. With this in mind a captured male was spared for mating with a normally colored female in the expectation that some of the litter would be white and a fine source of income. Alas, the cubs were all ordinary Bengals. Their sire seemed destined to cover the floor of the palace when the Maharajah was informed by his more knowing advisers that to get the benefit of the creature's recessive white gene, he had only to breed the tiger with one of its daughters. This was done, and sure enough, under the Mendelian law, the next litter included one "sport" along with three normal specimens. From this stock came Mohini Rewa, whom Dr. Reed cheerfully journeyed from Washington to India to obtain. "It's difficult to say how much the zoo owes to that cat and his cubs," he has said. "They drew attention to the facility and made all our recent improvements so much easier."

There was a drawback, however, and an instructive one. In mating several generations of daughters to a sire or a brother, the breeders were selecting deliberately for color, but unwittingly for less desirable qualities as well. The recessive genes had unattractive aspects which were simultaneously passed along. In time the descendants of the original white beauty became increasingly swaybacked, cross-eyed, and inclined to kidney trouble.*

It must be conceded that the sort of deterioration that befell the white tigers of Rewa is not an inevitable consequence of inbreeding; the pattern varies. Nevertheless, the general theory is that closely inbred animals will in time show both a dwindling vitality and a high rate of infant mortality. Accordingly, zoos must discard—and the better ones have done so—the old idea that their obligation is to offer the public a "postage stamp" collection of individual animals, that is, a few specimens of as many different animals as they can squeeze onto the property.

Reducing the number of *species* and increasing the number of *specimens* is now almost a badge of pride in the enlightened zoo. R. E. Honegger told me enthusiastically that the Zürich Zoo, where he is curator, disposed of valuable animals simply to enable it to concentrate on others with which it had already had considerable success. It gave away, for example, a full-grown pair of gorillas to make more room for its orangutans, and got rid of its Watusi cattle in favor of more blackbucks, its Malayan tapirs to get on with a second breeding group of tree kangaroos, and two species of crocodile, including a gift specimen from Fidel Castro, in order to specialize in the American alligator. Similarly, the Philadelphia Zoo traded, loaned, or sold all its antelope species except for three it wanted to concentrate on—the springbok, eland, and bushbuck. The National Zoo now has only half the primate species it once had and plans to eliminate all its chimpanzees, but it has a dozen groups of marmosets (forty or fifty individuals), a species on which it is doing

*The Maharajah, incidentally, continued to sell his "sports" to zoos and circuses until the government of India nationalized the animals, taking, as Warren Thomas, director of the Los Angeles Zoo, puts it, "the tiger's share of the profits."

intensive study, and ten groups of lesser pandas. Where the Bronx Zoo formerly exhibited 2,900 animals of 1,100 species, it now shows 3,500 animals of only 600 species.

THE SUBTLETIES OF MATING

Leading zoo people will readily admit that they are only beginning to know all that needs to be known about breeding wild animals in captivity. What they do know now is that each species requires particular conditions for mating, and for most of them these requirements and preferences call for considerably more than quartering a male and a female in the same enclosure and hoping for the best.

For many years this was the routine approach, and it failed more often than it succeeded. It was the regular course with cheetahs, for example, but no cheetah young were born in a zoo until 1960. It now appears, after much observation in the wild, that the species simply prefers the solitary life to companionship, and when a lone pair are thrown together over a period of time there is no breeding except of the contempt that comes with familiarity. Given adequate space for the loner's life, however, and the introduction of a male just at the time when he can do the most good, a female cheetah will mate as readily as any of the other cats. Some authorities think she is especially receptive if a second male is on hand to stimulate her interest with a bit of a contest, but success in this circumstance may just be a matter of having increased the mathematical chances, rather than of competition.

If experts still disagree on that point, they do agree on the need for a female cheetah's isolation over the long haul. A scientific report on the subject cites the case of one that had spent four years in an enclosure with a male. During that time the pair not only failed to reproduce; worse still, the female showed signs of stress, such as continual pacing and trouble in urinating, which indicated, the authors of the report thought, that "her territorial behavior was inhibited" by the unnaturally constant association with the male.

In contrast to the isolation required by the cheetah, some animals need long engagements, as it were. Writing of the Indian rhinoceroses at the National Zoo, Devra Kleiman reports that "successful copulation finally occurred after they were given one and one-half months of constant contact with frequent interaction." She adds that, in general, "problems in breeding many large mammals may be due to insufficient tactile contact between a pair."

Herd animals—some zebras, antelopes, seals, and many others—operate of course on the harem system, with one male doing all the servicing in a large herd and tolerating not even the presence of other adult males, much less their competition. If the younger or weaker males are to remain alive and sound, space has to be found for them as a bachelor herd elsewhere in the zoo. This is particularly true of certain antelopes, which fall into the behavioral patterns of their species regardless of the area allotted to them. Zebras differ from one species to another as to space requirements. The common zebras of the plains graze close together in large herds, completely tolerant of company. But put Hartmann's mountain zebras, notoriously aggressive loners, in a paddock too small for them to keep their distance from each other and, as one general curator explained, "they will kick the hell out of each other."

Among the primates there are great variations in the grouping essential to well-being—and therefore to breeding. Gorillas do well in an extended family arrangement and poorly just in pairs. According to James Doherty, general curator at the Bronx Zoo, "They need peer relationships almost more than the maternal relationship." Baboons, too, do best in troops, but the monogamous gibbon prefers the nuclear family —a pair and their young. The promiscuous chimpanzee does best in a fluid group, which males can join or leave without any apparent bearing on the breeding success of the troop as a whole. Among wolves the alpha (dominant) female controls mating, with only one female, usually herself, producing in the pack at any given time.

Beyond such distinctions among species, many a particular animal's mating habits become known only by long observation or simply by trial and error. They may be ingrained in a species or they may only depend, as with humans, on individual preference. The Los Angeles Zoo still has, at this writing, a female monkey-eating eagle (now known as the Philippine eagle) whose dislikes are notoriously violent. Paired with a male brought from the Bronx Zoo in the hope of increasing the population of this extremely rare species, she went beyond spurning his advances. When he tried once too often to push her toward the specially prepared nest, she simply killed him—a crime, it seemed, less of passion than of annoyance. The day I toured the zoo in the company of the chief animal keeper, Ed Alonso, she seemed determined to do him in for good measure, rushing at the wire fence of her huge aviary as though to club him with her huge, violently flapping wings. "A lovely bird," he kept saying in evident admiration of her spirit.

With some species of antelope, the male is the source of danger. At the Marwell Zoo it was found that at certain seasons the impala ram can be extremely aggressive and if left in the common paddock may kill a female or two, to say nothing of the calves. Paired tigers need particularly large enclosures to allow either mate to beat a quick retreat after copulation, when one participant or the other, usually the female, is apt to take a few murderous swipes at a recently amorous partner. Bears and rhinoceroses sometimes display this same misplaced ardor.

Each animal, it seems, has its own rules of the mating game, and any would-be breeder has to learn them. Officials of the National Zoo had to find out for themselves that their bald eagles would not mate so long as other birds shared their enclosure but did so readily when installed in a large flight cage of their own. Even within a herd system, a pair of sable antelopes have been known to refrain from copulating without privacy. A curator of the Frankfurt Zoo strongly suspects that some animals—zebras and clouded leopards are two examples—have a taboo against sibling sex, and it has happened that even a pair of unrelated gorillas who have grown up together will not mate in maturity because of their pseudo-sibling connection. On the other hand, incestuous couplings are quite the conventional thing for lovebirds.

Strangely enough, a stumbling block in the groupings of captive animals is the difficulty that often occurs in determining the sex of an individual to begin with. Zoo inmates are known to have been hopefully paired off for a year before it was discovered that they were of the same sex. The problem is particularly acute with birds, since 30 per cent of all avian species are monomorphic—that is, the two sexes are indistinguishable by size, plumage, or color.

Surgical methods of telling the sexes apart have been available for some time, and so have chromosome studies, but although these are not difficult, the stress induced by restraint can be harmful to easily traumatized specimens. The risk is particularly undesirable, obviously, with endangered species, such as the Puerto Rican parrot. This bird presented especially difficult breeding problems until the research department of the San Diego Zoo worked out the technique of measuring bird droppings for estrogen and testosterone, the ratio of those hormones being a reliable indicator of sex. The zoo's laboratory immediately planned to make its new fecal steroid analysis service available to other bird breeders, especially those with programs for raising rare species.

The sexing problem is not confined to birds. Giant tortoises are indistinguishable in this respect until they reach maturity, at fifteen to twenty years of age. By then it may be too late to obtain an individual of the opposite sex in order to launch a breeding program for these rare animals. Many reptiles are in the same difficult category, but fortunately the hormone test may be used even for such rare creatures as the Komodo dragon. A cellular test is sometimes used for the spotted hyena, long an object of derision among some African tribes because male and female are equipped with external genitalia that appear confusingly similar. The test involves coaxing the animal to the fence of its enclosure and plucking a few hairs from the hide. Cells from the hair root are then studied and generally yield the required information.

Sex and social grouping are only the first factors, of course, in putting zoo animals together for breeding. Another fundamental one is timing. The cheetah's case is just one example of the care required in deciding when to bring the sexes of certain species together. The poorest zoo can breed lions. Given a chance, they will mate in Times Square, and to such effect that their progeny are as hard to give away as domestic kittens. But the breeding of many other animals requires extensive knowledge.

There are great variations to begin with in the time that a species may take to attain sexual maturity, and then in the length of time it retains its potency before going into a sharp sexual decline. Devra Kleiman of the National Zoo explains that in the wild, 80 per cent of certain animal species, particularly some of the rodents, die before they are sexually developed. But for these species as a whole, nature compensates with large and frequent litters. At the other extreme, big animals such as elephants, whales, and the primates produce their young only at long intervals—five or six years in the case of elephants—and then one at a time.

Dr. Kleiman cites the mistake of a colleague at her own establishment who missed the boat with a collection of tamarins. These monkeys propagate very young and "with a bang," but their productive span is extremely short. In this case sexual senility set in so early that by the time the zoo was ready to bring pairs together, the males' interest in the activity had waned to the point of causing general disappointment.

LIVING SPACE

Besides allowing for natural groupings, a zoo's housing arrangements must take into account a factor about which little was known until after World War II, when Dr. H. Hediger, a top authority on animal psychology, worked out his rule of flight distances. It is a concept that bears significantly on another aspect of animal satisfaction in zoos. Given the predator–prey relationship that has existed for eons in the wild, every animal knows instinctively how close it can allow an enemy to come before taking to its heels. This flight or escape distance differs for every species, with size an important factor. Dr. Hediger explains:

> As a rule small species of animals have a short escape distance, large animals a long one. The wall lizard can be approached to within a couple of yards before it takes to flight, but a crocodile makes off at fifty. The sparrow hops about unconcerned almost under our feet, thus like the mouse, having a very short flight distance, while crows and eagles, deer and chamois for instance, have much longer ones.

An even greater variant than an animal's size in this connection is the particular environment in which it finds itself. In a part of the wild that happens to be a hunting area, flight distances will be greater than they are in protected reserves. Birds like the African superb starling and the weaver bird that I encountered as I walked in the safer parts of the Ngorongoro Crater allowed me to approach within inches, because they do not associate man with hunting. In the same way, as animals in captivity become used to humans and lose their fear of them, their flight distances sometimes disappear altogether.* In any event they shrink considerably in the security of an enclosure, which is a major reason why wild animals require much less space in a zoo than they do in the wild. The sometimes difficult question is to determine where the line is, for each animal, between tolerably less and harmfully less.

A considerable difference between hoofed animals and carnivores prevails in this respect, as an official of the Arizona–Sonora Desert Museum took pains to explain to me. A zebra or antelope, having a greater built-in fear of predators to begin with, may bolt at a sudden noise and run into a wire fence, which it won't even see in its panic to get away, or it may try to jump over a low wall. A certain amount of just such injury to hoofed animals does occur in zoos; but obviously the bigger the paddock, the less likelihood of injury.

---

*Or seem to. Actually the animals have learned to depend on the barrier separating them from zoo visitors. Relaxed by this protection, they will come as close as they can to a human being, having apparently no flight distance at all. But a visitor who foolishly climbs over the fence or wall will find that the zoo creature, lacking space to maintain its original flight distance, will almost certainly attack.

Large carnivores have greater escape distances than the hoofed animals, as a rule, and even though these shrink in captivity, there are limits. I recall seeing in a wretched zoo, some years ago, a magnificent wolf confined in a cage not much larger than itself. That is indefensibly cruel by any standard, but the kind of wolf wood that is a feature of many modern zoos, usually a long rectangular area with a good scattering of trees and brush, is not in the least cruel even though it is a far cry from the scores of miles that a wolf might travel in the wild—if it has to.

This condition is important. David Hancocks, director of the Seattle Zoo, points out that the wolf easily adapts to smaller space even in nature, and is doubtless glad to be able to. In the far north, where the pickings are slim, a pack may have to cover a territory of hundreds of square miles; but farther south, where more prey is available, it will satisfy its wants in a much smaller area. It should be emphasized that animals, unlike humans, travel strictly from necessity—for food, shelter, and climate—not for the pleasure of touring. It was a keen disappointment to the designers of the Busch Gardens that a long track provided for cheetahs to indulge a supposed fondness for swift running went practically unused. Given all the food they wanted, the cheetahs simply lolled about. Unlike Jogging Man, they saw no point in needlessly expending all that energy.

On this same question of cage size, it is worth noting that Dacca, a featured tiger resident in the Bronx Zoo for twenty years, spent her life in an indoor cage measuring approximately 15 by 16 feet and an outdoor cage measuring 20 by 25. In her remarkably long life in the Bronx, Dacca bred thirty-two healthy cubs and displayed no neurotic behavior whatever. Dr. Conway, who is anything but a defender of the old-fashioned zoo cage, concedes that some tigers have done as well as Dacca in even smaller enclosures, whereas others have been a lot less satisfied in cages considerably larger. There is clearly more to sound housing in the zoo than mere size.

At least as important is the scope provided in the enclosure for the tenant to follow the habits of its kind. To an astonishing degree, animals in the wild are conservative in their ways, in fact compulsive. Every activity in the life of a wild creature may have a fixed spot in its territory —a place to sleep and be safe, a place to eat, a place to defecate, a place to give birth and hide its young. The territory of an animal may have "marking" points that it uses to define its area by leaving feces, urine, or glandular excrescences; and generally all these fixed points are connected by paths well worn by countless trips over unvarying routes.

So far as it is possible, zoo animals must be allowed to follow a similar pattern of existence, as many of them make an effort to do. The size of an enclosure may not be as important, therefore, as the provision it makes for at least some of the functional points and facilities around which the inmate's life revolves—burrows for some, wallows for others; nesting boxes, scent posts, and comparable objects or arrangements, each depending on the particular needs of the animal as observed in nature.

Dr. Hediger, whom it is impossible to avoid quoting in any discussion of animal behavior, tells of a remarkable experience in this connection. Having noticed in Africa that zebras regularly rubbed themselves, to the great improvement of their coats, against the hard termites' nests that are common throughout their habitat, he decided to duplicate those structures in the zoo that he directed in Zürich. The moment the stable door was opened and his charges saw the new object (made of a cement mix) in their paddock, they proceeded to rub against it so vigorously that it soon toppled over. A reinforced version was quickly put up, and so eager were the zebras to get at it that two attendants had to hold them off with whips until the synthetic termites' nest had sufficiently hardened. It has been a daily source of pleasure ever since, as it is now in zoos all over the world. Zürich also has a wooden post for rhinoceroses to rub their horns against. And in Frankfurt, keepers have thoughtfully placed a large rock in the center of the lion enclosure to allow the subdominant males of the pride a hiding place for escaping the watchful eye of the boss lion.

The number of nesting boxes or denning sites in an enclosure can be of major importance. A pair of tree shrews, it was discovered at the National Zoo, will sleep in one nest and keep their brood in another. If forced to share sleeping quarters, the couple can get annoyed enough, if that is the right word, to take the rather extreme step of eating their young. A similar regard for the natural habits of bird species has long since paid off spectacularly at the Bronx Zoo. When well-simulated habitats for waterbirds were established there in 1964, breeding immediately occurred among species which in the old-fashioned bird cages had not produced an egg in half a century.

In nature the rockfowl builds its nest on a small projection beneath an overhanging cliff. When one at the Frankfurt Zoo was seen trying to build on a vertical wall, just below the ceiling, keepers quickly put up a small platform to oblige it. Nests, appropriate perches, and nesting

Hsing-Hsing, the National Zoo's male giant panda, peers out
from behind his morning meal of bamboo.
(Photo by Ilene Berg, National Zoological Park, Washington.)

Margay and kitten—forest cats
easily confused with ocelots,
found from Mexico to Brazil.
(Photo by Stephen M. Alden,
Arizona–Sonora Desert Museum.)

Frosty, the Bronx Zoo's polar bear cub, being "licked into shape."
(Photo by Bill Meng, New York Zoological Society.)

Family gathering of giraffes in the Serengeti-like freedom
of the Busch Gardens, Tampa.
(Photo by David Hume Kennerly, Busch Gardens, Tampa.)

TOP: Trio of Przewalski horses, extinct in the wild but not at the Rotterdam Zoo.
(Photo by Diergaarde Blijdorp, Rotterdam Zoo, Rotterdam.)

BOTTOM: Manchurian cranes, not yet endangered but getting there, at the Rotterdam Zoo.
(Photo by Diergaarde Blijdorp, Rotterdam Zoo, Rotterdam.)

Bearded dragon, a gentle Australian creature
in spite of its fierce look and thorny scales—
this one at Chicago's Lincoln Park Zoo.
(Photo by Bud Bertog, Chicago Park District.)

OPPOSITE TOP: Père David deer, harem-master
with harem—a Chinese species extinct in
the wild for centuries.
(Photo by Ilene Berg,
National Zoological Park, Washington.)

OPPOSITE BOTTOM: White-handed gibbon enjoying
a gymnastic life on Bronx Zoo's Gibbon Island.
(Photo by New York Zoological Society.)

The Bronx Zoo's Przewalski horses,
closest thing to the Mongolian mounts ridden
by the hordes of Genghis Khan.
(Photo by New York Zoological Society.)

An all but invisible moat keeps African predators from their natural prey at the Milwaukee County Zoo. *(Photo by Milwaukee County Zoo.)*

Playful river otter viewed through glass panel
at the Arizona–Sonora Desert Museum.
(Photo by Arizona–Sonora Desert Museum.)

Female Philippine, or monkey-eating, eagle, fierce enough
to have frustrated the Los Angeles Zoo's breeding effort
by killing her would-be mate.
(Photo by Neal Johnston, Los Angeles Zoo.)

Kangaroo and young at
the Rotterdam Zoo.
*(Photo by Diergaarde Blijdorp,
Rotterdam Zoo, Rotterdam.)*

Bactrian (two-humped camel),
now considered vulnerable in its
native Asian wild,
flourishing at the Washington Zoo's
breeding park at Front Royal Virginia.
*(Photo by Ilene Berg,
National Zoological Park, Washington.)*

One of the Washington Zoo's
famous tiger "sports,"
with white fur, chocolate brown stripes,
and blue eyes.
*(Photo by Ilene Berg,
National Zoological Park, Washington.)*

Newly-hatched bog turtle, rapidly losing
its wetland habitat in the northeastern
United States and rarely found in captivity.
*(Photo by Bill Meng,
New York Zoological Society.)*

African Plains in the Bronx, featuring natural predator and prey, artfully separated. (*Photo by New York Zoological Society.*)

Fred, a lowland gorilla happily ensconced
on a durable post at Chicago's Lincoln Park Zoo.
(Photo by Bud Bertog, Lincoln Park Zoo.)

Bighorn sheep in its natural setting at the
Arizona–Sonora Desert Museum near Tucson.
(Photo by Arizona–Sonora Desert Museum.)

Mexican wolf pups, endangered in the wild but
doing well at the Arizona–Sonora Desert Museum.
(Photo by Arizona–Sonora Desert Museum.)

material are made available at all good zoos now, and some put net curtaining over the aviary's glass to keep fledglings from flying into it, in the manner of the migrating birds that sometimes crash into the glass walls of skyscrapers.

Not least in a zoo's concern with breeding its charges is the need of some animals for privacy. Occasional refuge from human observation is essential to a great many species, and a modern zoo will take that need into account even if the visitor finds it a bit more difficult to see a particular creature the minute it strikes his fancy. A new exhibit in the Los Angeles Zoo allows its wolves so much room to hide that at times no animal is to be seen. The wolves like the arrangement, but it could create a problem in the zoo's public relations. So far, however, curators have taken the reasonable position that it is good for visitors to expend a little time and effort to locate their "quarry," to a small degree as though they were in the wild and having to summon up a degree of patience before they merit the reward of seeing a fine specimen.

So vital was privacy to a snow leopard recently come to motherhood in the Bronx Zoo that her keepers went to extraordinary lengths to prevent any human intrusion. To give Shanda a feeling of complete security while she nursed her young, they closed the lion house entirely for several weeks. The disappointment of visitors was forestalled by giving them access to a television monitor that showed more of the intimate scene than could possibly have been observed in the building, although the primary purpose of the screening was to enable keepers, rather than visitors, to keep an eye on how mother and cubs were doing.

SCIENCE AS MIDWIFE

Despite all such care and precautions, however, situations arise in zoos that call for more direct human intervention if breeding is to succeed. The most dramatic form this intervention can take is the collecting and storage of sperm plus artificial insemination, even now an established procedure but one at an extremely early stage of development. Dr. George Rabb, director of the Brookfield Zoo in Chicago, plays it down for the present: "A lot of it is still just talk." But, he adds, it is "good talk," meaning it is more than a dreamy hypothesis.

Techniques are already well established. Not only has the procedure been used with domestic animals for many years, but even in zoos it has at least proved its feasibility. Young wolves, peregrine falcons, golden eagles, and gazelles are a few of the additions that zoos have acquired as a result of artificial insemination. Sperm has been successfully drawn from lions, tigers, polar bears, giant pandas, and wild horses, and for some time now the London Zoo has been storing elephant semen in much the same way that cattle sperm is stored as a common practice.

The sober judgment of zoo authorities, however, is that artificial insemination in zoos will not be widely or readily practiced in the near future. I found Reed and Conway of an identical mind regarding lack of knowledge as the barrier. "Of course it is done every day with cows," Conway concedes, "but cows are thoroughly familiar and all the same." Species differ enormously, especially in timing, and it will take years of trial and error to discover the best moment and the best way to extract semen from males of varying structure, to find the optimum conditions for storing it, and, hardest of all, to discover for each species, the optimum time to inject it into the female. Techniques for the first step include manual masturbation of the male (now done with Andean condors at the Bronx Zoo)—no easy matter, obviously, even with some sedation—the use of artificial vaginas for collecting sperm, and electro-ejaculation, a method obvious from the name.

Insemination calls for detailed knowledge of the female's reproductive processes, which vary greatly from animal to animal. With minks and rabbits, for example, ovulation is promoted by the violence of the male's sex behavior—a rabbit can be made to ovulate by picking her up and shaking her—and the rough treatment of the female in some big cat species prior to copulation may serve the same purpose. No one is quite sure, but if this sort of thing is an essential part of the picture, then clearly the artificial insemination of tigers calls for courage above and beyond the call.

The period in which fertilization can occur likewise varies greatly from species to species and in some cases may be no longer than twenty-four hours out of the year. The trouble is that this strategic span is not always detectable. "Take the giant panda," says Dr. Reed. "We know when she comes into full cycle but we have to watch her like a hawk as she builds up to it; with other animals, we don't know at all when they ovulate."

Dr. Conway echoes these same doubts and adds another: artificial insemination is more expensive as a continuing procedure than it would be to keep sound breeding groups of the relevant animals to begin with. "You have to keep track of cycles, do the extracting and inseminating at just the right time, and maintain the equipment necessary for storing the semen." This last aspect of the procedure alone would require nitrogen tanks, additional personnel expert in refrigeration, and much more. All of which can put a strain on zoo budgets, which are rarely robust to begin with.

On balance, artificial insemination is still regarded as "an increasingly useful technique with a definite role in the future, but no substitute now for maintaining viable herds." Even if its increasing usefulness is gradual, however, the technique is sure to be tried more and more in cases where a male animal, such as a bull elephant, is hard to transport and hard to keep; where a rare animal such as the giant panda is, for reasons yet unknown, difficult to breed naturally; and in situations where new blood is needed to forestall inbreeding in a well-established herd. If the procedure can in time be made to serve in these ways, the rewards will be great even if one discounts science-fiction scenarios of a future in which creatures then long gone from the earth will be re-created by combining sperm and eggs from the deep freeze or by cloning from long-stored genetic material.

On a more modest level of human intervention in the breeding process is the zoo-keeper's service as role model in the raising of zoo progeny. Where a young primate mother has not herself had the benefit of parental or peer models in maternal behavior, this intervention can be vital to the survival of the young. All too often it has happened that a new gorilla mother has no idea what to do with her baby when it is born. In nature, she would have served her apprenticeship as an "aunt," learning as part of her life in the group how to handle and care for infants. With her own first-born in a zoo, she may be baffled and fearful, likely to reject the baby at the very least and sometimes hostile enough to kill it. But the loss of a new gorilla being intolerable—on economic as well as humane grounds—zoo personnel will go to great lengths to train the mother in advance of the great moment.

At the San Diego Zoo a film was tried for this purpose, but two-dimensional figures without smell seemed only to confuse the prospective mother, a lowland gorilla named Dolly. A surrogate baby was then introduced in the form of a stuffed home-made doll, with the face drawn in ink. This seemed more acceptable than film and was made even more so by the soft repetition, over two months, of simple commands—"Pick up the baby, Dolly," "Be nice to the baby, Dolly," and so on—along with demonstrations by the trainer, Steven Joines, of how those acts were to be performed. As reported by Mr. Joines in a paper in *International Zoo Yearbook*, for 1977, the patient training paid off magnificently when Dolly, who had rejected her first-born, immediately picked up her second, took it to her breast, cradled it, and otherwise performed the duties of her office. The same experiment was tried soon afterward, with equal success, at the Seattle Zoo.

In most aviaries "double clutching" is a simple device used by zoo managers to stimulate the growth of rare bird populations. As soon as a clutch is laid, it is removed by the keeper for hatching in an incubator. The immediate effect of the removal, with a majority of species, is to stimulate the bird to lay a replacement, which may in turn be removed. Dr. Conway reports that individual cranes at the Bronx Zoo have been induced in this way to lay as many as six clutches in place of one, and an Andean condor, which normally lays one egg every two years, recently quintupled its production of chicks through use of this devious but harmless device.

Sometimes animals are given hormones to stimulate the reproductive process, but at the Rotterdam Zoo officials expressed some caution on this score. The offspring of such animals, I was told, might in turn require hormones before they could reproduce—a circumstance that would make impossible their propagation in the event they were ever returned to the wild. The zoo people in Rotterdam obviously had an eye on the eventual reintroduction of some animals to their native habitats.

A negative intervention in breeding zoo animals, but one of the most important, is the prevention, with certain exceptions, of cross-breeding. The soundness of that objective is not questioned by any reputable zoo official today. Most of them look with scorn on such perpetrations of the past as mating a tiger and lion to get a crowd-pleasing, if infertile, "tiglon." That abomination could not occur in nature and even in captivity only in circumstances of total boredom and the absence of normal sexual outlets.*

The exceptions are another matter. Where a zoo has, say, one male orangutan of the Borneo subspecies and one female of the Sumatran subspecies, until recently it had been generally considered permissible to cross them for the sake of propagating the endangered orangutan species itself. The same might be true of two subspecies of tiger. But even this is severely frowned on in some zoos, now that other means of promoting species propagation are available. At the Wildlife Preservation Trust, Gerald Durrell's famous small-animal establishment on the Isle of Jersey, the prevailing view is that "as conservationists we must

*Not long ago a gibbon and a siamang mated by chance in the Atlanta Zoo and produced the first ape hybrid, in spite of marked differences in the chromosome arrangements of the two animals. It remains to be seen whether the young ape will itself prove fertile.

confine our efforts to maintaining the characteristics of the animals that have evolved over thousands of years isolated from each other. If ever reintroduction appears viable, only the local variety should be considered, except when failure to do otherwise might result in the loss of the whole species."

In the light of this exception, then, there are situations where an argument is still made by some zoo managers for a very limited degree of crossing between closely related animals. Taking the long view, Dr. Hediger is more inclined than his colleagues to minimize the case for purity. If we still have orangutans a century from now, he says, it won't matter whether they are Bornean, Sumatran, or a mix. Only narrowly specialized zoologists can identify twenty-eight different subspecies or races of African rhinoceros, "and to distinguish one from another even they would have to kill the animals and study the skeletons."

Hybrids aside, no human intervention in the breeding process in zoos is more basic than the introduction of fresh blood into an established group—but even that is far from simple. We have seen that animals have not only their species-dictated requirements for mating, but also their individual preferences. There is no assurance that a new male zebra, for example, will be acceptable to the females of a particular zoo herd, or that he will take to his new surroundings with enough security to let him fulfill his mission.

The expectation that this same zebra will in fact prove a good sire rests on further changes that have come over the best of our zoos—in addition, that is, to the startling advances already touched on. The first of these—part of an interzoo relationship which was unknown a few decades ago—is the ease with which collections now lend each other animals for the purpose of breeding, usually with an agreement to divide the progeny—"first male for us, first female for you" (or vice versa).

On the East Coast, the Bronx and National zoos have developed a smooth working relationship of this sort. "Conway's got the sable antelopes and we've got the Père David deer," says Theodore Reed. "We're already having joint strategies for these animals and more zoos will come into the picture."

Similarly, on the West Coast, Los Angeles sends its female rhinoceroses to San Diego, which has a proven stud, whereas the Los Angeles male has yet to prove dependable. In England, the London Zoo, its affiliate at Whipsnade, and Marwell in Hampshire jointly own and raise a herd of the rare Hartmann's mountain zebra.

TRADE AND TRAVEL

Fortunately advances in transportation now allow more distant interzoo traffic of this sort. The pink-cheeked ibis, one of the rarest of bird species, is being saved by a series of breeding loans that went first from Basel and Tel Aviv to the Isle of Jersey and then, as offspring arrived, from Jersey to Philadelphia. San Diego has a female okapi by courtesy of the Cheyenne Mountain Zoo in Colorado Springs. And National has sent its Galápagos turtles to Hawaii, where climate and soil promise greater breeding success.

Top zoo people realize that if species are to be saved, such interzoo cooperation still has a way to go from the competitive spirit of the old days. "There's a lot of trading going on—breeding exchanges and the like," says Dr. Reed, "but managing scimitar-horned oryx as a national unit in order to preserve the species and mix up the bloodlines—well, not everyone is at that stage of the game yet." Until they are, he and other leaders in the field intend to keep a check on the welfare of animals that they send out. Before letting one go, they satisfy themselves not only about its genetic prospects in its new home but also about all other conditions maintained at the recipient institution. It is one way to raise the standards of zoos.

Finally, among the improvements forced on the zoos of the world by the need to breed is a far greater emphasis on safety in transport. There is no use in arranging to borrow a snow leopard of proven quality as a stud if he is to be so traumatized on the way that his services will not be worth his travel ticket.

In the annals of the London Zoo, opened in 1827, there are accounts of how animals were transported from the wild to Regent's Park in those early days. The zoo's first hippopotamus, having been captured for Queen Victoria by a detachment of Nubian soldiers, was floated down the Nile on a barge, like Cleopatra, and nourished on milk from a flock of goats brought along for the purpose. The zoo's first chimpanzee arrived in a stagecoach, and its first giraffes, walking through the streets from the dock, the zoo reports, "panicked at the sight of a cow in Commercial Road." However quaint, for many specimens such travel was also lethal. Journeys from all parts of the world took weeks or months by ship, and a great many specimens succumbed to stress, inadequate food, accidents, and other hardships of the journey.

It was a long time before procedures substantially improved. In a request for a federal permit to import four female lowland gorillas in

1972, Warren Thomas, then the director of the Gladys Porter Zoo in Brownsville, Texas, conceded to the Bureau of Sport Fisheries and Wildlife the past hazards of such an undertaking. One of these, he wrote, was the high incidence of disease right down to the period following World War II:

> *Since milk and other nutrients were rarely available, it was a common practice to hire native wet nurses in an attempt to keep the gorilla babies alive. More often than not, this met with disaster because of a condition called "thrush," which is a mycotic infection passed through the milk. This condition was harbored by nearly 90 per cent of the native women of West Africa.*

No wonder so few gorillas—and even fewer breeding pairs—wound up in zoos.

What has made all the difference in animal transportation is the airplane. The exceedingly rare okapi, unseen by a white man until this century, was totally unable to survive the long trip to Europe from what was then the Belgian Congo—by river boat, train, and ship. Of ten brought out of the jungle in 1949, eight died en route and one soon after its arrival in Basel. Thanks to the speed of air transport and the easy control over such factors as temperature, small okapi populations were soon established in the zoos of the world. Their numbers have increased to about sixty specimens, half of which were born in captivity. Unfortunately, that number is still far too small to give much assurance that the zoos and the okapis' native forest combined can save this strange animal, which belongs to a genus of its own although it resembles the short-necked giraffe that became extinct some ten million years ago.

Rare animals traveling today—from zoo to zoo, generally, since they can no longer be taken from the wild—arrive by air or truck, in appropriate boxes, crates, stalls, bags, or tanks. Giraffes are transported, as a rule, only when they are still young so that special routing is not required to avoid low overpasses. With all species, in fact, young specimens are favored since they adapt more readily to change than their elders.

It might be supposed that the tranquilizing gun would come into play for travel, as it does for medical treatment, but usually it does not. "We try not to use sedation for animals in transit," says Gerald Lentz, director of the Busch Gardens, "because we wouldn't be able to tell then how the animal was taking the trip or what physical requirements it might have en route." It is better to use holding barns for a few days —that is, narrow stalls resembling the container to be used in transit— in order to accustom the animal to confinement and see how it reacts. For the more excitable specimens, blindfolds may be in order.

It is true that for some animals travel, especially to a drastically different climate, can still produce a fatal shock. Bird losses, especially, may be as high as fifteen per cent either in transit or in quarantine—a fact often cited in condemning zoos. But what is not generally conceded is that not more than ten per cent of a zoo's birds are likely to be imported in the first place, so that as far as zoos go, we are talking about a loss not of fifteen per cent but of only 1.5 percent—that is, fifteen per cent of the imported ten per cent. The figures for bird transport fatalities cited by critics of zoos do not apply to them so much as to pet shops and dealers, whose trade accounts for 98.5 per cent of the fatalities. This is so primarily because zoos are generally careful to deal with reputable collectors, but also because they have occasion to order only one or two birds of a species at a time, whereas the dealers take them in crates of scores or even hundreds. Plane flights of short duration, and knowledge-able crating, give the Lincoln Park Zoo in Chicago the satisfaction of not having lost a single bird in transit over a period of four years.

THE BREEDING RECORD

With behaviorists and biologists steadily learning more about the group-ing, sex habits, and housing needs of wild animals in captivity, what success have the zoos actually been having that might point to them as the future trustees of the world's wildlife?

To deal with this question, fairly enough raised by skeptical critics, the obvious place to begin is with the zoos' most dramatic claim, the one that their opponents concede even while some of them minimize it. Sim-ply stated, it is that species are being bred in zoos today that would otherwise even now be altogether extinct. Saved from that fate are the following:

Père David deer. Sometimes known as the *milu,* this is an unattrac-tive but highly interesting creature whose Chinese name, *sse-pu-hsiang,* means "not like four"—not some prescient reference to the Gang of Four but a description of an animal that has attributes of a stag, a goat, an ass, and a cow, yet manages on the whole to resemble none of them. It has a long tufted tail, strangely shaped antlers, and large splayed hooves—these last perhaps an adaptation to the marshy areas it presum-ably once inhabited.

In any event, the deer was discovered in 1866 by Father Armand David, a French missionary who located a herd in the Imperial Hunting Park near Peking. Although he could not have known this at the time, no specimen had existed in the wild for well over two thousand years. Those in the emperor's deer park were fated to be wiped out in their turn during the hectic days of the Boxer Rebellion, except for one specimen that survived until the early 1920s.

Fortunately some of the animals had been removed in the decades following Père David's visit, and of these the Duke of Bedford rounded up eighteen from European zoos for breeding on his estate. In spite of losses in the hard days of the two world wars, the herd flourished well enough, in what has since been converted to a wild animal park, for specimens to be redistributed to zoos in Europe and the United States, and recently in China itself. The *International Zoo Yearbook* census records a population close to eight hundred in seventy-six collections, all descended from the Duke's herd of eighteen.

The Przewalski horse. Last of the true wild horses, this handsome yellowish-brown animal, with its short, Prussianlike mane, is thought to have been the carrier of Genghis Khan's hordes on their forays across China and Russia. Hunting and competitive grazing had long been driving these equines, named for the Russian explorer who rediscovered them, into territory with inadequate water, but even so their end in the wild came rather suddenly. In the 1940s there were still sizable herds, and a number of the animals were caught, with difficulty, and brought to European and American zoos. That was fortunate. By 1960 they had practically vanished from the Mongolian plains, and it is all but certain that none now remain in the wild. But as the studbook kept at Prague shows, the number in captivity has quadrupled since then, though the herds are considerably inbred. They are concentrated in the Catskill Game Farm in upstate New York, Askaniya–Nova (U.S.S.R.), Prague, Marwell, Whipsnade, York, and San Diego.

All told, there were 254 by last count, and they had lost none of their wild ways, even though their blood appears to have been somewhat diluted by a cross somewhere along the line with that of another equine, probably a Mongolian or Norwegian pony. The choice has been between preserving an imperceptibly changed Przewalski horse (and even animals in the wild change somewhat over a period of time) and having no such horse at all.

The wisent or European bison. Hunting and the steady clearing of Europe's wooded lands had by 1800 driven this majestic subject of Cro-Magnon art to the confines of a few Polish–Russian forests. The five hundred or so remaining specimens a century later were virtually exterminated in the wasting savagery of World War I, the last one being left to the slight mercies of a poacher in 1921. Happily more than fifty had been acquired by zoos and animal parks before that occasion, and an International Association for the Preservation of the European Bison had been founded in Berlin.

With good diet, adequate space, and proper grouping, the Berlin herd flourished. Bulls were sent to other European collections and a studbook established to keep careful track of the animals in order to assure breeding lines as vigorous as possible. In the destruction of World War II, the wisent—not among the war's most memorable victims, to be sure—had its population once more gravely reduced. This time Swedish zoos kept the species going, distributing breeding pairs to other collections in the decade that followed. Thanks entirely to their work, the wisent came back enough to allow a small herd to be reintroduced into the Bielowecza Forest. Here and in nearby Soviet forests, it is now slowly returning to a natural existence.

True, the wisent's habitat today is something of a reserve, where supplemental feedings are provided when needed and a general watch is maintained. But the species *has* been saved, and it has been saved by zoos.

It should be noted here that the American buffalo, close cousin to the wisent, is a species kept alive in a similarly controlled setting. It too owes its existence to dedicated conservationists, operating under the aegis of the American Bison Society. Down from 60 million to a remnant of three hundred at the turn of the century, the buffalo was carefully bred, most notably at the Bronx Zoo, and its herds were built up until they could be farmed out to bison ranges and national parks in the West, where they now number more than 35,000.

Birds belonging in this category of salvation-through-captivity are the Hawaiian goose, the Laysan teal, and the Swinhoe pheasant. But it is not so much these few creatures, all of which had already reached the brink, that principally support the zoos' claim to glory for their rescue operations, but rather the dozens of species they are managing even now to keep away from the brink.

Consider just the primates that, thanks to the efforts of zoos, are already moving away from certain oblivion toward at least some measure of numerical restoration. Difficult as the great apes are to breed, since a female generally produces only four or five offspring in her lifetime, the zoos have already run up an impressive record. With the number of orangutans in the wild estimated at hardly more than 2,500, according to the *International Zoo Yearbook*'s 1976 rare animal census there were 648 in zoos, of which 279 had been bred in captivity. Rotterdam, Philadelphia, and several others have had second-generation orangs, a true test of breeding success. A decade ago only four or five orangutan births a year could be expected in the world's zoos, but by 1975 the number had risen to 42 such births, with 34 surviving. These numbers are still not high enough for complacency, but an official of the San Diego Zoo assured me unequivocally that orangutans should be able to continue indefinitely in zoos, "even after they are gone in the wild." By agreement among the zoos themselves, incidentally, no more of these apes will be imported even under special allowances made for captive breeding.

The zoo production of chimpanzees is roughly double that of orangutans, but the gorilla's future in captivity is less promising. With fewer than a thousand left in the African wild, including both lowland and mountain varieties, there were 457 reported in captivity in 1978. Unfortunately only about 22 per cent of these were zoo-born. With no further supply from the wild probable, it will take considerable doing to keep the species going. Still, the 1978 figures compare favorably with the count taken twelve years earlier, when there were 170 gorillas in zoos, only six of which were born there. Chicago's Lincoln Park Zoo alone now has sixteen in its splendid group, and even the charming little private zoo at Wassenaar in Holland has five of the glorious animals, (not to mention two rare bonobos, or pigmy chimpanzees). Although that may be too many for a small establishment, it can be argued that the gorillas are better off there than in a place where they may end up having their skulls sold as souvenirs—as still happens in their ancestral countries in spite of earnest efforts by government to stop the practice.

As for lemurs, zoos may already have made the difference between early extinction and the saving of a species—in fact, of several species. Having persuaded the Malagasy (Madagascar) Government to let it have a few couples of each lemur subspecies to breed before their numbers had shrunk to impossible levels, the San Diego Zoo embarked on a highly successful conservation program. In eight years it had a population of twenty-one black and white ruffed lemurs, eight red ruffed (which were down to something like thirty in nature), and sixty-two of the much more common ring-tailed variety.

So great had been the success of the program that some of the progeny of the group originally taken from the forest were returned to

Madagascar. Altogether, zoos now hold close to a hundred ruffed lemurs. Indeed, the species' ultimate chance of survival appears to rest with zoos, where it propagates quite readily, rather than in its land of origin, now that over 90 per cent of their forest habitat has disappeared.

Similarly stringent efforts are now being made, especially at the National Zoo, to save the golden lion tamarin (much like a marmoset), but the prospects are still doubtful—as they are in the coastal forests of its native Brazil. Elsewhere efforts are being made on behalf of the lion-tailed macaque, a gray-whiskered monkey also known as the "wanderoo," which is about to vanish from its native Indian homeland.

The Asian one-horned rhinoceros, down to about 450 in the sanctuaries of Nepal and Assam, at last count numbered 61 in captivity, of which 28 were born in zoos. Two of these awesomely implausible creatures were sent to the Basel Zoo in 1951–52 and made the subject of careful observation and study. The reward has been ten offspring, some of which were in turn paired off at other zoos. It is thrilling to see these prehistoric animals in the Kaziranga animal preserve, as I did; but it is good to know that if that fine effort eventually gives way to human—or inhuman—pressures, there will still be a good possibility of seeing them in Basel, Amsterdam, Hamburg, London (Whipsnade), Stuttgart, West Berlin, Washington, and the Bronx. The white rhinoceros (not white at all but called that from a mistaken translation of the Boer word, *wijde*, referring to its wide, square-lipped mouth) is alive and well in South Africa today, thanks almost wholly to zoos and animal reserves.

Among the cats, the case of the tiger is outstanding. Of the 450 to 500 Siberian tigers now in zoos (there are fewer than that in the wild) some 90 per cent were born in captivity; of the pure Bengals to be found in zoos (500 to 600), something over 80 per cent. If these figures do not reflect enough of a triumph of productivity over consumption in zoos, there is Dr. Conway's additional statistic that all the 1,144 Siberian tigers ever counted in the international stud book (including those now dead) came from an original stock of 39 wild-born specimens.

In 1972 no more than 5.7 per cent of zoo cheetahs were captive-born, but the figure has since risen steadily as zoos learned the animal's breeding habits. Only six years later the percentage was up to 31, and it is most likely higher than that by now. Whipsnade alone records 57 cheetah births since 1967, including some third-generation offspring. Figures for the much rarer snow leopard are even more startling. In 1964, only 16 per cent of the specimens to be found in zoos had been born there; a decade later, with the population up from 49 to 161, more than two-thirds were zoo-born.

With hoofed animals the success of breeding programs has become great enough to create a delicate and growing problem. The natural grouping for many of these animals—deer, antelope, some zebras, and wild asses, among others—is a harem, in which only one adult male can be accommodated in a zoo compound. As the population increases, there are accordingly more and more *surplus* males, and although several of these extras can easily be held in reserve as a literal "bull pen" for genetic variety and emergency service, space for a large bachelor herd, such as one finds in nature, can become costly in acreage and impractical as an exhibit. Yet there is no other way these animals can be kept with any fidelity to nature.

Although lions, tigers, and other carnivores are not herd animals, they, too, are beginning to present a surplus-male problem, one or two being sufficient for an adequate breeding program. Proper housing for a larger number can be forbiddingly expensive—not to mention the feed bill, with meat for a full-grown specimen coming to something like $2,000 a year.

Obviously the first thought of a zoo manager with an extra male on his hands is to sell it, lend it out to smaller zoos for breeding, or simply give it away. But in a great many cases, no other collection wants another male; or, if it does, the would-be recipient may not be regarded as a fit guardian. The excellent Rotterdam Zoo, I was told, simply will not dispose of an animal to an "undesirable customer, like a safari park or a circus." Moreover, for an animal rare in nature, the red tape involved in shipping it across state or national lines may be too formidable to warrant the effort.

For some time various population controls have been used in zoos to ease this problem. The male lions at Brookfield, for example, have both had vasectomies. All Rotterdam's lions and tigers are kept on the birth control pill except when the zoo has a request from another collection for a cub, in which event the best pair available are put on a breeding program. It is possible, of course, simply to keep the sexes apart when the female is in estrus; but with some species it may be very difficult to tell, just from observation, when an individual female is coming into heat. Finally, all devices that prevent the normal behavior of captive animals draw objections from some zoo managers on the ground that they detract from the educational value that such behavior has for the zoo-going public—and from the well-being of the animals as well. "Castration, sterilization, all these things," says Dr. Peter Weilenmann, director of the Zürich Zoo, "change the animal," and even contraception may affect its chances of becoming a parent in the future.

Inevitably directors and curators have come to talk—so far, more often in Europe than in the United States—of "recycling," a euphemism for using euthanized animals—aging, surplus, and indisposable—as food for their carnivores. I found Dr. Faust, in Frankfurt, entirely open on the subject. He had already had to shoot some sable antelope, he said, for lack of space, and he preferred to destroy them rather than subject them to "very bad lives" in safari parks of the kind that might have taken them off his hands. "But the time has not come," he concluded, "when we might have to do the same with tigers."

At the London Zoo the term is "judicious culling." In order to "maintain young and viable stocks," reads a recent *Scientific Report* of its Zoological Society, "over 130 head of stock, mostly ungulates of non-endangered species, were culled in the period under review" (1975–77). Although the practice has not yet become prevalent in this country, many top zoo managers see it coming. Dr. Reed was, as always, forthright. Some day, he said, he might have to stand up in front of his environmentalist friends and say something like, "Come to my house, we are having Père David deerburgers." After all, he added more seriously, "We have to worry about the species, not the individual. We can't return those deer to the swamps"—where, indeed, no Père David deer has lived for centuries.

It remained for William Conway to put the entire question in perspective, as he did at the 1976 annual conference of the American Association of Zoological Parks and Aquariums:

> *First of all, zoos can take pride in the fact that the surplus problem has arisen because zoos are getting better . . . Although zoo populations lack many of the checks and balances of animal groups in nature, all healthy wild animal populations produce more young than are needed or could be accommodated within normal breeding populations . . .*

To take care of these,

> *zoos still must face the problem of removing animals which are superannuated, diseased, abnormal or simply too many of one sex, age, or genetic line, the animals that would have been lost in*

a wild population through predation, disease or other natural misfortune . . .

Zoo men have not been able to deal as calculatedly with their collections as, for example, farmers do with their livestock or as humane organizations do with dogs and cats. Thirteen and a half million surplus dogs and cats were put to death by United States humane organizations last year; about nineteen times as many animals as exist in all the zoos in the world . . .

Despite our own emotional involvement with our animals, when no other humane and philosophically suitable alternative is available, zoos must not shrink from seeing that surplus animals are mercifully destroyed . . .

Groups devoted to humane treatment of animals might generally be expected to register automatic outrage, but Sue Pressman, director of wildlife protection for the Humane Society of the United States, is wholly sympathetic. As one with years of experience as supervisor of health in three zoos, Mrs. Pressman knows too much about the subject to oppose euthanasia—"not just for the old and sick, but also for surplus animals" that undermine the well-being of a collection. As in nature everywhere else, she observes, "death is a fact of life," and where a zoo-keeper is too sentimental to meet his responsibility to his collection in this way, she would personally step in and "help him out."

Over all, close to 150 of the 280 mammals listed by the IUCN as endangered, rare, or vulnerable are now to be seen in zoos. Moreover, approximately a hundred of these species and subspecies are represented by specimens bred in captivity—although, according to an authoritative study in 1976, only thirteen of these species could yet be considered truly self-sustaining in zoos. The thirteen are the white-throated wallaby, lion-tailed macaque, Bengal tiger, Siberian tiger, Sumatran tiger, leopard, Przewalski horse, swamp deer, Père David's deer, lechwe, scimitar-horned oryx, addax, and Arabian gazelle. As of 1980 it is probably fair to add the chimpanzee and orangutan, and a number of others, if not yet self-sustaining in zoos, give good indications of being headed that way—among them the black lemur, golden lion tamarin, ocelot, timber wolf, clouded leopard, Asian lion, cheetah, banteng, and pygmy hippopotamus. Unfortunately, it cannot be said that this degree of success has yet been reached for birds, reptiles, or amphibians.

Some critics insist that it is not zoos as such that should be credited with the extent to which the prospects for animals have improved, but rather the spacious animal parks and so-called "survival centers" where propagation is the major objective. This seems hardly a valid distinction, however, in view of the fact that these same parks, farms, and breeding centers are in almost all instances adjuncts of zoos, established and operated by them and designed to serve their needs as well as to add to the general wildlife population. A glance at these remarkable facilities may indeed convince the skeptic of the good zoo's seriousness of purpose and the major contribution it can make in the years ahead.

Few of those who visit the Bronx Zoo can have any idea of the enterprise it operates for propagating rare animals on St. Catherine's Island, off the coast of Georgia. On that narrow barrier island, about twelve miles long, is the New York Zoological Society's Rare Animal Survival Center. A half hour by boat from the mainland and closed to the public except by permission, St. Catherine's combines savanna grassland, salt marshes, and tidal flats, not to mention a native fauna that includes alligators, ospreys, sea turtles, and bald eagles. To these have been progressively added animals either hard to breed in a zoo or requiring more space than most zoos find available.

Starting experimentally in 1975 with a pilot colony of gemsbok, an unendangered species of antelope, the Center's pastures and corrals have become home to such other ungulates as the threatened or endangered slender-horned gazelle, sable antelope, dama gazelle, addax, and sitatunga, as well as the red kangaroo and Grèvy zebra. Among the rare birds that take naturally to the mild Georgian climate are Hawaiian geese, Indian sarus cranes, wattled cranes, and such rare representatives of the parrot family as the red-fronted and hyacinth macaws and the palm and Leadbeater's cockatoos.

The island's new fauna have been thriving in spite of a rocky start because of infections from feral cattle, which have since been removed. The enterprise is made possible by the support of the Edward John Noble Foundation, but its operation is in the hands of the Society, that is, the Bronx Zoo. Cooperation is the order of the day, and breeding animals have been sent there from the National Zoo, the San Diego Wild Animal Park, the Busch Gardens, and the Philadelphia and Denver zoos. St. Catherine's, in turn, has already sent some of its surplus gemsbok and gazelles elsewhere.

In the near future, according to John Lukas, the associate curator, the Center plans to take on endangered sea tortoises, lemurs, and condors. "We will be able to supply animals to zoos, whose sources in the wild are being cut off more and more every year," he said. "Here we can provide space, a good climate, and without the public to think about we can study our animals in a scientific way, with all the genetic factors in mind."

On an even larger scale, thanks to the federal status of the Smithsonian Institution, is the National Zoo's Conservation and Research Center near Front Royal, Virginia. Here, bordering the Shenandoah Valley, Eld's deer from Burma, scimitar-horned oryx, two-humped Bactrian camels, and zebras roam at will in the gigantic fenced pastures of what was once a United States Cavalry breeding center and later, among other things, a camp for German prisoners of war. On the great grassy stretches, European bison are at home. Smaller but still spacious areas accommodate aggressive Persian onagers, and there are wooded retreats for herds of Père David deer.

For ungulates, the advantages of space at Front Royal are obvious. The huge size of the paddocks makes it possible for a few males besides the dominant one to stay with a herd and even occasionally to sire some of the young, assuring a greater genetic variability than is possible in the one-male zoo compound.

Besides the hoofed stock, Front Royal has tree kangaroos, golden marmosets, and lesser pandas—little red creatures which are as appealing, if not as greatly endangered, as their giant namesakes. A particularly interesting group are three rare South American canids—the bushdog, the crab-eating fox, and the maned wolf. Closed-circuit television has been used to observe the social organization and habits of these animals, very little being known about how best to handle them in captivity. Endangered birds bred at Front Royal include, among others, Rothschild's mynah, a handsome white bird with a masklike patch of blue on the head, the sandhill crane, and the Grand Cayman Amazon parrot.

A staff of only twenty operates this 3,000-acre facility, with a few additional employes taken on to harvest the hay and alfalfa in season. But scientists and students are generally on hand to make those painstaking observations that add invaluably to our knowledge of exotic creatures and that few can ever hope to make in the wild.

Other major breeding centers, not as well endowed financially as St. Catherine's and Front Royal, are obliged to share some of their terrain with the public, whose admission fees make possible the work done largely behind the scenes. Notable among these is the San Diego Wild Animal Park under the same management as the San Diego Zoo, but some thirty miles north of the city. On 1,800 acres, encircled by a monorail for somewhat distant viewing, are herds of hoofed animals—sixty specimens in the addax herd alone. "That would be unthinkable in the typical collection," says the park's general curator, James Dolan. "They might have three or four at the most." In the ideal grouping made possible by adequate space, the park has had a record crop of species less than common: Hartmann's mountain zebra, Arabian oryx, slender-horned gazelle (at least thirty), cheetah, Formosan sika deer, and bara-singha. It has recorded the birth of a lowland gorilla and can boast more white rhinos than any place in the world outside of South Africa. All this at the expense of having to maintain such attractions—happily confined to one end of the park—as "Nairobi Village," complete with a "Fishing Camp" and a children's "Petting Kraal."*

Busch Gardens, in Tampa, Florida, goes quite a bit further in the direction of flamboyant entertainment, as it is probably obliged to do in order to sustain a $7.75 charge at the gate, but here, too, the end result is impressive. Well away from its magic shows, jugglers, snake charmers, and belly dancers, all happily clustered near the entrance, is a fine animal collection, well exhibited and easily viewed from monorail or skytrain. Perhaps even more important, Busch Gardens has an extremely good breeding record for rare species. Under the guidance of Gerald Lentz, a serious and experienced zoo man, it has produced more than forty specimens of each of sixteen species of hoofed animals, including addax, scimitar-horned oryx, and Grèvy zebra, as well as Soemmering's gazelle, Hunter's hartebeest, and roan antelope. The collection is also rich in birds, such as the maribou stork, that are not generally found in zoos.

For breeding loans and a chance to purchase odd ungulates, many American collections are indebted to one of the more unusual animal enterprises, the Catskill Game Farm. From this privately owned park about a hundred miles north of New York City—and its connected ranch at Ocala, Florida—have come hundreds of the hoofed animals, rare and not so rare, that are featured in the country's zoos. Both the farm and the ranch are the privately owned and managed operations of Roland Lindemann, one of the most respected of American animal experts.

Not a professional scientist, Lindemann makes a distinction between the theoretical zoologist that his father was and the expert in

*Perhaps it need not be too apologetic, however, because if the village were really to resemble Nairobi—such is the pace of change in Africa—it would have to have a Hilton hotel and a Southern fried chicken stand on the main thoroughfare.

animal husbandry he himself became. He might well have bred horses for the owners of racing thoroughbreds instead of making himself a shrewd raiser of antelopes of all species, an outstanding dealer in zebras, vicuñas, and giraffes. But from the start, his husbandry took the scientific approach. He was among the first animal handlers to experiment with tranquilizers and antibiotics, and few of the professionals know more about the chromosome differences that separate a given species from a closely related one. "There is not a zoo in the country," he says with a disarming forthrightness "that did not at one time or another have a representative at Catskill to see what we've done, how we do it, how we treat our animals."

Entranced as a youth by a pair of Przewalski horses in the Berlin Zoo, Lindemann was one day to gather the specimens hidden away on European farms during World War II in order to raise them in this country. Sixty-four of the surviving specimens were bred in the United States under his management, and of these thirty-six were then distributed to various collections. Almost every Przewalski horse in America, I have been told, originated from the stock he collected.

The Catskill Farm is open to the public for a charge that helps pay for the upkeep, but most of the income is from trading. When I interviewed him at his Ocala ranch, Mr. Lindemann had to leave suddenly on an unexpected trip to arrange the sale of a pair of giraffes to Nelson Rockefeller—whose death a week or two later put an end to plans he was reported to have entertained for establishing a wild animal ranch of his own.

Combining the features of the Catskill trading–breeding–displaying operation with the survival center motif of Front Royal is Marwell Zoological Park, a few miles from Winchester, in Hampshire. Here are some of the biggest herds of desert antelope in captivity, particularly the scimitar-horned oryx, and of the Przewalski horse as well. The park, founded in 1972 by John Knowles, another highly regarded unacademic breeder of wild animals, has also been extraordinarily successful with Siberian tigers. Marwell carries on a heavily patronized trade and breeding-loan operation throughout Europe. Its delightful monthly *Zoo Paper* combines the features of a stock report, a maternity ward journal, and a once-over-lightly gazette about doings in the world of zoos.

Thirty miles from London, in the rolling parkland of Bedfordshire, is Whipsnade, one of the greatest of these combination zoo-and-propagation centers in the world. Founded by the London Zoological Society in 1931 as a kind of rest home and sanitarium for its temporarily ill or disabled animals, Whipsnade gradually developed into a magnificent country zoo, which now specializes in conservation and in supplying other collections with the offspring of its breeding programs.

These efforts have been so successful that over 80 per cent of all the mammals to be seen at Whipsnade were born and raised there. None have been imported from the wild, I was told, in thirty years. Principally a backup for the London Zoo—and a place to get around the surplus problem with large bachelor herds—the park has excelled in Père David deer, white rhinos, and musk oxen, as well as rare Chilean flamingos, polar bears, and pygmy hippos. Thomson's gazelles, which do not always do well in zoos, have thrived at Whipsnade even to the extent of developing coats thick enough to withstand the English weather. But for the casual visitor, perhaps nothing more becomes the place than the hundreds of small creatures that have the complete run of the park. Peahens may be seen roosting on the low branches of trees, and at any turn you may spot a muntjac deer no bigger than a dog, or a little Chinese water deer with a wispy moustache, or a colony of prairie dogs, or a brightly plumaged Indian jungle fowl, or a wallaby bounding over a meadow.

To single out these great centers for successful breeding is by no means to minimize the productive achievements of many of the urban zoos themselves—quite apart from the intimate connection that some enjoy with these same centers. If one were to consider only a few in order to make the point, one would do well to start with the awesome record of the Basel Zoo, affectionately known to generations of Swiss as "Zolli" (a diminutive for *Zoologische*), whose collection includes second-generation Indian rhinos, colobus monkeys, and gorillas, third-generation pygmy hippos, and fourth-generation tigers. "Zolli" has likewise done extremely well with Malayan tapirs, echidnas, spectacled bears, proboscis monkeys, kookaburra birds, and wild asses—all rare or endangered.

The Frankfurt Zoo was the first to send orangutans back to Sumatra and chimpanzees back to Tanzania. It pioneered in the regular breeding of all four species of ape, including the rare bonobo or pygmy chimpanzee; it appears to hold the record for success with the rapidly disappearing okapi, and has bred no fewer than seventeen gerenuks, gazelles of an uncommon species that stand on their hind legs to nibble at foliage.

So it goes—and so it should go, since the degree to which zoos specialize is the degree to which they can breed, distribute to other zoos, and thereby build up the populations of threatened species. At Zürich the specialties are the snow leopard, the vicuña, and the fish otter, which was once on the brink of extinction in Switzerland. Los Angeles is big on gorillas, bongos, and mountain sheep. And the Durrell collection on the island of Jersey has been the salvation of a remarkable variety of such little-known animals as the tenrec, a Madagascan version of the hedge-hog; the Jamaican hutia, a West Indian rodent; Goeldi's monkey, a rare product of the South American rain forest; and the even rarer waldrapp, an ibis with a curved beak, iridescent plumage, and practically no chance for survival in the wild unless its North African and Middle Eastern habitats can eventually be restocked with captive birds from this haven in the English Channel.

The Importance of Creature Comforts

My years of observing animals from the other side of the bars have convinced me that most of my guests live longer, healthier, and happier lives in the zoo than in the wild.

—WILLIAM MANN

For zoo animals to thrive, something more than appropriate housing, natural social grouping, and the necessary conditions for mating, vital as they are, is needed to ensure the general well-being that alone can compensate for captivity. The recognition of this need to keep its charges in good health and spirits has produced further extensive change in the zoo as an institution and made fresh demands on its personnel—starting with the requirement to provide each individual animal with the right food in the right quantity and in the form most appealing to it, if to no other creature on earth.

Chinese scientists have recently concluded, to cite an extreme example of fastidious tastes, that the giant panda is dying out in the mountain forests of Szechwan because its preferred diet is so limited. Now believed to number fewer than a thousand, these attractive animals depend in winter almost exclusively on two species of bamboo which, as nature would have it, bloom once in a century. Thereupon the plants drop their seeds and die off, leaving pandas without adequate sustenance over the long stretch that it takes for a new bamboo crop to grow from seed to edible stalk.*

The panda, unfortunately for zoos, is not the only fussy eater in the animal kingdom. The koala subsists entirely on eucalyptus leaves. What makes things difficult, according to Grzimek, is that koalas of one Australian area may reject a particular species of eucalyptus that is wholly acceptable to those of another. Worse still, there is excellent reason for such fussiness: one koala's chosen leaf is literally another's poison. All of which adds up to the fact that a zoo curator has to know both his koala and his eucalyptus—or risk the life of a rare and charming creature.

These are extreme cases, admittedly, but the range of food requirements from species to species—not to mention the variation of taste from individual to individual—is enough to stagger anyone who ever supposed that in zoos the herbivores live simply on hay and the carnivores on horse meat. Knowing the nutritional needs of their varying charges is not enough if curators can neither locate the native food nor discover a palatable substitute. The National Zoo is fortunate in being able to supply its two giant pandas with their weekly 350 pounds of bamboo, mostly by cultivating the tree right in its own precincts but also by accepting regular donations from Washingtonians who grow it ornamentally in their backyards.

Since the ways of animals in the wild were not closely studied until recent decades, the diet of zoo inhabitants for many years had to be worked out by trial and error, and the findings passed along from one institution to another—at least where the fierce rivalry that once marked interzoo relations did not preclude such cooperation. The starting point, says John F. Eisenberg of the National Zoo, was to use domestic animals as models—dogs for the canine family, cats for the felines, horses for the zebras, and so on. Possibly on the theory that gorillas are close to man, one of the first of these great apes brought to Germany a century ago is reliably reported to have breakfasted on such local staples as Hamburg smoked meat, frankfurters, and cheese, washed down with "a glass of white beer with a fruit flavor."

Although it is true that some carnivorous animals in zoos will take to any kind of meat, the problem of feeding them does not properly end

*The Peking Zoo is reported to have found chicken-and-rice broth acceptable to giant pandas, and the National Zoo in Washington adds fruit and vegetables to their diet, along with sweet potatoes, dog biscuit, and assorted vitamins. But all these are side dishes and no substitute for bamboo.

there. In nature, the Bronx Zoo's James Doherty explains, they get a balanced diet by eating their entire prey—stomach, liver, kidneys, heart, tripe, and all—each organ, plus the bones, satisfying a specific nutritional need. That is very different from getting a slab of red horse meat, "which is like ice cream for them"—something they like, "but not at all a balanced meal." For that they now get, at the very least, additives that supply the necessary vitamins and minerals. Beyond these general diets, there are species and individual variations. Tigers and jaguars, for example, want more fat than lions do.

Meat-eaters generally prefer live food, but it is rare in zoos that they can indulge themselves in that taste. An exception might be made for a reptile or small mammal that is old or ill and will not eat otherwise, but overwhelmingly zoo animals are prohibited from practicing the natural cruelties of life in the wild—a fact often overlooked by sentimentalists whose sympathy for a caged lion does not extend to the antelopes he might have eaten on his home grounds. Nor, inconsistently, does it extend to the crickets that go, all unmourned, down the gullets of zoo-dwelling frogs, or to the grasshoppers that tempt the palates of newly hatched crocodiles. Chicks and mice by the thousand are consumed in zoos, but the only animal allowed to kill them there is man.

Zoos will in fact go to considerable length to substitute for live prey, although I was told at the Basel Zoo in Switzerland that an occasional live rodent is offered to the bigger snakes while the keepers are still trying to interest them in less objectionable fare. At the Gladys Porter Zoo in Brownsville, Texas, snakes are given sausage that has been wrapped in a cloth and left for a while to absorb the odors of an empty rodent-breeding unit so as to give it the acceptable flavor. A king cobra may even be given its sausage served up in the skin of some smaller snake that happens to be an item in the cobra's natural diet. The deception makes it possible not only to rule out live feeding but also to avoid the constant hunt for small snakes, accompanied as they generally are by a regiment of parasites.

An idea of the complexity of zoo diets for birds of prey is suggested by the comment of a British veterinary pathologist in a paper contributed to the *International Zoo Yearbook* for 1977: "Ideally, aviary birds should be fed on a diet of laboratory rodents; these are likely to be free of infection, low in pesticides and balanced nutritionally. Day-old chicks are popular with falconers but a thiamine deficiency has been associated with their use . . . and they probably only offer marginal levels of protein and minerals." The Bronx Zoo not long ago had a special problem with the blood it had been feeding its vampire bats. Routinely discarded by hospital blood banks after three weeks, this nutrient did well enough until instances of hepatitis increased to the point where the zoo found it safer to feed its bats on beef blood.

Zoo-goers are sometimes disappointed or critical at seeing a paddock for grazing animals that is something less than a rich green, or one for browsers that has very little in it in the way of live trees. A second's thought will of course make it obvious that neither grass nor foliage would survive more than a few weeks of the cropping and munching that such species naturally engage in. Hay, some branches, and perhaps hydroponic grass grown for the purpose—these and pellets are their real source of nourishment.

Composite foods were pioneered at the Philadelphia Zoo in the 1930s, and since then their use, in one degree or another, has become worldwide. Dr. Herbert L. Ratcliffe, director of the Penrose Research Laboratory, connected with both the zoo and the University of Pennsylvania, devised the famous "monkey cake" as well as diet formulas for many other creatures to be found in the zoo. Good for all omnivorous mammals that will eat it—palatibility is sometimes a problem—the monkey compound includes cereals (corn, wheat, barley, or oats), plant proteins (peanut or soybean meal), animal protein (skimmed milk powder), and minerals (calcium phosphate, salt, *et al.*)—to which vitamins are now regularly added. The recipe typically adapted by the Basel Zoo, as reported in the *International Zoo Yearbook* for 1977, reads: "Mix nine parts of the basic mixture with one part minced cooked meat, combining with meat broth or water to make a stiff mash. Press into a shallow pan. The mash hardens quickly to form a cake, which can be easily cut or broken into pieces for feeding. Keep in refrigerator." A serving may be anywhere from a twentieth to half an ounce for each pound of the animal's weight, depending on how active it is. Twisting the human order around somewhat, Basel offers its young apes and monkeys a dessert of vegetables and fruits—only if they have first eaten their cake.

As a basic diet, such concentrates have considerably simplified the life of zoo chefs and, as the Philadelphia Zoo has shown over the past forty-five years, they have promoted the health and longevity of their patrons. Massa, the lowland gorilla, celebrated his fiftieth birthday in December 1980—an attainment twice the life expectancy of gorillas in the wild—and two of its orangutans died at fifty-six and fifty-seven, having broken the record for captive primates. Some of the collection's birds have done similarly well as senior citizens.

Yet there is still some range of opinion concerning packaged foods. At two major zoos I visited, pellets were only sparingly given to ruminants, which need a certain amount of browse in their diet. Pellets can be "overzealously" used, in the opinion of Dr. Eisenberg, who prefers to offer any new animal—especially such small, esoteric ones as the colobus monkeys, langurs, and some marsupials—a "smorgasbord" of feeds based on what is known, however little, of its eating habits in the wild. "Hit the Safeway," he suggests, "and buy out every green in stock, letting the animals gradually select out."

Dr. Poglayen, at the Arizona–Sonora Desert Museum, thinks pellets wrong for the equids—zebras, wild asses, and the like. She says hay is the thing for them, mostly timothy. Since they would quickly strip down any grass, their paddocks are planted to vegetation they don't care for. In her particular zoo, yucca fills the bill quite nicely because zebras will not touch the stuff. At Whipsnade I found the curator, V. J. A. Manton, even more forcefully opposed to any standardizing of zoo diets. "I'm anti-pellet," he says flatly, explaining that no animal in the wild eats the same thing day after day or season after season. "We don't feed by rote, but by the individual judgment of the keepers."

Most zoos do find packaged goods invaluable in degrees varying for different species, but rarely as an exclusive diet; even Manton favors pellets "as a booster." They are widely used along with fruits, vegetables, branches, and, especially in city zoos, hydroponic grass. The Bronx grows enough of this carpetlike green to use 1,500 pounds a day, finding it "fresher than anything we could buy"—and considerably cheaper, no doubt, than lettuce.

The following daily menus for just two animals should convey some idea of the comprehensiveness of zoo cuisine:

GORILLA (NATIONAL ZOO)
1½ lb. kale
1 lb. meat, cooked
2 scoops monkey chow
3 slices bread, ¼ lb. sweet potato
5 apples, 5 bananas, 4 oranges
1 egg, cooked
  Once a week add ½ lb. carrots, ¼ lb. onions

ELEPHANT (LONDON ZOO)
100 lb. hay
2 lb. oats
2 lb. maize

4 lb. locust beans
4 lb. biscuits
10 lb. carrots
10 lb. potatoes
3 or 4 cabbages
1 loaf bread
  Apples, oranges, 1 oz. salt, sometimes fresh leaves, bamboo
  shoots, dried fruit, cod liver oil

Special animals, of course, require special foods, and one can only hint at the scores of instances—aside from koalas and giant pandas—where devoted curators and keepers take extra pains to cater to the appetites of their charges. From the maned wolf's broad molars, experts at the Frankfurt Zoo realized that plant food made up part of its natural diet and accordingly added fruit and oats to the meat naturally served up to an essentially carnivorous animal. The delicate pygmy chimps at San Diego get, besides their commercially made monkey-biscuit, a dish of Gerber's high protein baby cereal with milk and brown sugar, additional milk, and a variety of fruit.

Many zoos issue a daily quota of fresh leaves and branches to all their herbivores in the summer, thawing out, in winter, similar foliage that has been stored and frozen for the purpose. For two baby bears the National Zoo once provided weekly desserts of rice pudding with raisins, brown sugar, cinnamon, salt, and maple syrup, which the cook occasionally dipped into. And the snow leopards at the Helsinki Zoo get chunks of beef or lean pork into which the kitchen has carefully injected appropriate quantities of calcium, green meal, dried yeast, and skimmed milk, as well as vitamins A and D. On that diet, Helsinki has been remarkably successful in breeding this highly endangered cat, with a record of forty-two cubs since 1967.

None of these menus, whether regular or special, is at all out of the ordinary. They have been cited merely to show the degree of knowledge and care that now go into the feeding of zoo animals. These accordingly thrive to a degree unapproached by their kin in the wild, the vast majority of which suffer disease, undernourishment, or injury in the endless battle to survive. Says the naturalist Roger Caras, "Animals live twice as long in zoos as they do in the wild." Whether or not one dismisses the contrast as smacking too much of man-playing-God, it can hardly be denied that the health of zoo animals is a major factor in calculating the chance of salvation for at least some of the earth's endangered species.

The complete security that captive animals enjoy can exact a price in boredom, at least among the more intelligent animals. Good zoos have come to understand the negative health effects of such boredom and are coping with it. Since most of an animal's activity in nature involves the pursuit of food, ingenuity is being applied to make eating in the zoo more than a matter of nutrition. Although primate appetites may very well be satisfied by monkey cake and other compounds, for example, the animals' spirits and behavior are greatly improved by giving them the opportunity of searching for some of their food as they do in nature. In almost all the zoos I visited, straw scattered about the floor of the enclosure contains buried tidbits that add interest to a primate's life. Between a day's meals, naps, climbing, swinging, and mutual grooming, monkeys and apes put in hours searching the floor cover for raisins, sunflower seeds, currants, puffed wheat, and even, occasionally, crickets. In some zoos the treasures are frequently changed in order to provide an element of surprise. This may not make for tidiness, but it does make for healthier animals.

Grazers and browsers, although they depend essentially on the hay and pellets they can count on, are still free to engage in their endless nibbling. They simply require far less grass and shrubbery and can munch on the limited quantity available more as a behavioral activity than as a source of nutrition. Even though no zoo can supply its giraffes with an endless supply of trees, only a little ingenuity is needed to satisfy them, perhaps with hay and acacia leaves suspended from the branches of a bare tree, or with leafy food put up in high baskets several times a day to keep the giraffes coming back for refills.

Imaginative touches of this sort are what one should look for in passing judgment on a zoo. It is not so much the size of an enclosure that matters—though with some species that obviously has some importance—as what is in the enclosure to engage the life of the occupant as it might be engaged in nature. David Hancocks, director of Seattle's Woodland Park Zoo, put it rather well: "It is the quality of space that matters, not the quantity. If you had to spend a weekend in a superdome without contact with other people, you would be going up the wall with boredom by Monday morning. But if I locked you in this office [a small one] for the weekend, and gave you a radio, books, pencils, and so forth, you would keep yourself occupied."

A stereo would have helped in my case, and that is precisely the point: it is vital to know what would help in the case of each animal. The dozing zoo lion that you indignantly imagine is being subjected to a life of unutterable boredom would look no more energetic when viewed in the Serengeti. There he would be sleeping twenty-one hours out of twenty-four just as in the scroungiest zoo. But when the lion in the wild is awake and alert, he can be very active indeed. He does have some social life in his pride—and, who knows, perhaps some pride in his social life. Hence the combination of an indoor cage where he can peacefully snooze away the hours is acceptable provided he also has access, along with other lions, to a much larger outdoor cage—or better still, a replicated African plain such as the best of zoos now provide. But it is a mistake to imagine that even in the average indoor cage he is decaying from ennui when he is merely doing what comes naturally to cats, to wit, snoozing.

Primates are a different matter. For them boredom is a direct consequence of a zoo's lack of ingenuity. Space for them does not have to be wide or deep; it has to be vertically generous and equipped with however many playthings, swings, bars, and jungle gyms as can be got into it without defeating the purpose, which is to allow the residents the freedom and opportunity to jump, swing by their forelimbs, or climb, according to their own mode of behavior. The gadgets need not be elaborate. I have seen small-fry primates in a zoo occupied for fifteen minutes at a time with putting paper bags on their heads and taking them off suddenly, like human infants playing peekaboo. The sticks, bags, and other small attractions may be changed every few days for variety. The more ingenious pattern of swinging chains in some zoos is altered periodically, in effect turning the apparatus into a new toy. Going even further, the Milwaukee Zoo from time to time moves its monkeys—and others of the less routinized animals as well—from one locale to another just to provide a change of scene.

To really appreciate the strides zoos are making in combating the boredom of their tenants, one should visit first the old primate house at Chicago's Lincoln Park Zoo, now in limited use, and then the new one. The former has cages of the traditional kind wholly inadequate to accommodate the most successful gorilla colony this side of Zaire and Rwanda. The new house features what appears to be a brightly colored telegraph pole, enabling a great ape to climb three stories, to find comfortable platforms en route, and to be observed through glass from different levels by visitors who climb the stairway alongside. Frankfurt's primates can take advantage of plastic flaps that allow them to decide for themselves whether they want to be in or out of doors, giving them not only that option but probably an additional pastime in maneuvering the panels.

Activities of one sort or another are not, of course, designed solely for primates. Although I did not see it, I was told at Basel that meat is sometimes hung on a rope and swung, just to keep tigers in trim and mindful of the leaps required in nature in order to procure a meal. Predators are often given large wooden balls to play with, and Hediger suggests that even birds can appreciate diverting objects, citing the fact that macaws like to play with stones.

Although keepers are generally encouraged to play with the animals in their care, those in most zoos have little time for that attractive aspect of their work. Nevertheless, a keeper for the great apes in the National Zoo is quoted in *A Zoo For All Seasons:* "I feel an important part of my job is to take care of them mentally as well as physically. If they want to play, I put down whatever I'm doing and indulge them."

In a larger way, skilled personnel are having more intensive contact with zoo animals than ever by way of training acts. These are still controversial, but the notion that animal training must be merely for the entertainment of visitors is being replaced by the belief that, properly conceived, the discipline involved can be educational for the zoo-goer, helpful to the keeper, and, above all, stimulating for the animal.

Until recently the prevailing view was that such shows are demeaning to the performers and in any event unnatural. David Hancocks still thinks so. "They distort reality and reinforce stereotypes, such as the idea that chimpanzees are clowns. There is enough excitement in what animals really do, if you can demonstrate it, so that you don't need all that nonsense." A great deal of significance, however, may be attached to that "if." No responsible zoo people whom I encountered would pretend there was any value in having a sea lion evoke from a series of horns a crude approximation of "Pop Goes the Weasel." But to have it leap into the air for a ball or stick does in fact demonstrate the animal's proclivity for throwing a fish into the air so as to catch and swallow it in just the right way.

Dr. Hediger, who calls this "disciplined play," makes out a strong case for it in his major work, *The Psychology and Behaviour of Animals in Zoos and Circuses,* as he did in the interview I had with him in Zürich. "Many animals turn stupid when shut up in cages and left to themselves. Healthy activity of the occupational therapy sort can be a real benefit to them." Freed from their chief activities in the wild— avoiding enemies and seeking food—they need to fill the vacuum or they will suffer the kind of boredom that, especially in badly run zoos, shows itself in pacing and other aberrant behavior. The "tricks" are simply an outlet, preferably an intelligent and appropriate one, for the animals' latent energy.

With adult zoo specimens of some species it is neither safe nor fruitful to try establishing a human–animal play relationship, but it is possible to teach them activities that evidently give them pleasure. Indian elephants at the Frankfurt Zoo are taught to carry out thirty-two different commands, and (evidently because they get satisfaction from doing so) they may sometimes be seen going through the performances by themselves, with no trainer to give the orders. Other zoos, particularly the San Diego Wild Animal Park, have similar programs to keep their elephants "alert and active," as they must be if they are to enjoy a life in captivity sufficiently to flourish and reproduce.

Some of this training, particularly of elephants, is incidentally useful in caring for them and is sometimes defended solely on that ground. Obviously, obedience to specific orders, such as to lift a leg, is helpful if keepers are to trim an elephant's toenails (which are trimmed in nature by a great deal more walking than is called for in the zoo), or to have their feet checked for stones and possible injuries, not to mention their tails for parasites.

As far as possible, all animal performances, called "behaviors" by perhaps overly sensitive zoo officials, are related to the ways of the animal in nature. Elephants in the wild do on occasion stand on their hind legs or kneel, so it is not misleading to display those unexpected abilities to zoo visitors. Similarly, the bird show at San Diego Wild Animal Park is more than mere entertainment, though it is that in part. Besides the mimics doing their usual thing, it is fascinating to see how the great horned owl's specialized feathers, as well as its acute sight and hearing, allow it to swoop down silently on its prey, or to observe the pigeon's homing instinct in action, or to admire the teachability of Jo-Jo the raven as he snatches a dollar bill from the hand of a willing contributor in the audience, delivers it to the trainer, and then returns it to the donor. The running commentary of the trainer, though hardly solemn, is intelligent and informative. "The only way we can justify having our birds," he will explain, "is to use them in educating the public, to try to teach people how valuable birds are in nature"—and then let them go.

There is no comparable intent to return trained dolphins to the sea, but the sleek beauty of their bodies as they leap to unbelievable heights in the Milwaukee Zoo (to take one good example), their evident relish in the organized play, and the demonstrations of their superior intelligence —these are not only a revelation to their audiences but a sure way to enlist their sympathy for the protection of creatures whose brains are as complex as man's.

In spite of everything that proper food and psychological health can be expected to do for zoo animals, they will at times inevitably suffer from diseases, injuries, and sometimes disabilities related to the conditions of captivity. A terrain slightly different from what a species is used to, for example, can have crippling consequences unless compensated for in some way. Even the possibility of catching respiratory illnesses may call for special measures to protect the zoo inmates from the germs of their observers.

No less than in the science of feeding zoo animals suitably, great strides have been made in veterinary medicine. Until the past decade a resident veterinarian was the rarest mammal to be found in a zoo, and even now there are reported to be no more than forty in all the zoos of the United States and Canada. Most institutions have veterinarians on call, of course, but the constant watching and reporting that are at the heart of preventive medicine for captive animals falls to the keepers. It is they who watch for such telltale signs as listlessness and lack of appetite, which indicate the presence of parasites and worms—a universal zoo problem kept under control only by constant observation and frequent examinations of fecal matter. Hediger suggests that even variations in the smell of an animal can tip off the alert keeper to illness, although he also concedes, "Not everyone responsible for animals goes so far in the interpretation of such scent signs as some experts on mice, who believe they can identify surprising variations in the internal state of their tiny charges through the change of body smells."

Among the serious diseases that zoo people have to watch for are polio, rabies, distemper, tetanus, and tuberculosis.* Vaccinations and other injections for all of these are routine, along with periodic boosters and tuberculin tests, plus periodic weighing for some animals. The fatal Newcastle's disease, to which birds are highly susceptible, is now under control as a result of federal regulations so strict that finding the virus in one bird at a quarantine station automatically dooms all the other birds then on the premises—a regulation that meant the loss, a few years ago, of ten healthy penguins on their way to the San Diego Zoo.

Every zoo has its own quarantine for new arrivals, but viruses and

---

*Not to mention less expected ailments, such as cirrhosis in cheetahs, hepatitis in chimpanzees, and tumors in many marsupials, as well as cancer in a range of species, including lions and elephants.

bacteria continue to defy that barrier, along with everything else that preventive medicine can do. Ironically, enlarged enclosures, which are hard to keep clean, and the grouping of animals in natural herds—both admittedly major advances in the psychological well-being of the animals and the education of zoo-goers—have increased the risk of epidemics that from time to time take their toll.* Protection of primates from human carriers of tuberculosis is a major factor in the now common use of glass as a separator from the public, and the need to ward off parasites still accounts in many zoos for those sterile-looking cages whose tiled walls and shiny metal equipment suggest well-cared-for public lavatories rather than the haunts of wild creatures.

The printed guide to Frankfurt Zoo, an institution which has been criticized by some people in the field for having a "clinical" look, explains the thinking that has dictated the choice of materials for its primate cages:

> *In this house both the climbing apparatus and the light steel netting are made of stainless steel. This smooth material can be cleaned easily. Apart from this, the poles are much easier for the monkeys to hold on to than the thick wooden beams which were used earlier. They permit far more gymnastic possibilities for the same space. As wood is a breeding ground for parasites, we have entirely avoided its use here. For this reason we have also made use of heated, non-porous synthetic flooring and plexiglass sliding doors . . . sight screens inside the cages help to provide refuge for animals of inferior rank.*

An understandable reaction has set in among zoo directors against these oversanitized enclosures, for which advances in drugs and medical technology have in fact reduced the need. But it does not follow that the physical attributes of animal housing are less important than they ever were to the health of the zoo population. On the contrary, sound zoo planning now requires the participation not only of architect and director, but of curator and veterinarian as well.

Special care has to go into the surfacing of enclosures for equids and antelopes if their hoofs are not to grow to inordinate length, a condition that can be as dangerous to the animal as high heels would be to a female jogger. When I visited the Philadelphia Zoo, a zebra was being darted

---

*An outbreak of anthrax at England's beautiful Chester Zoo in 1964 killed off all four of its elephants and eleven of its small carnivores. The elephants had to be dismembered before it was possible to cremate them, a procedure required by law in the circumstances.

so that a blacksmith could shave down hoofs which, in spite of efforts to reproduce a native terrain, had not been worn down as they would have been on the African plains. Sand is sometimes scattered on the floor of an animal's enclosure so as to increase the wear on its hoofs, as well as to preserve its muscle tone and also to keep it from slipping.

Knowing the habits and limitations of an animal is as important as the danger of contagion in deciding how to separate it from the viewing public. Snow leopards, because they are capable of great upward leaps, can hardly be kept in an open pit unless it is one so deep that visitors can get only a long-distance view. The giraffes at Whipsnade are in a paddock that slopes down to water. Because they are wary of losing their footing, zoo officials know there is no danger that they will attempt to ford the stream.

Some thirty years ago the Bronx Zoo lost M'koko, one of its most popular characters, in a drowning accident that resulted from a poorly planned moat. Gorillas do not swim, and when the magnificent 500-pound male inadvertently slipped into the moat surrounding his enclosure, the dropoff proved far too sharp. Failing to grab the cables that had purposely been left dangling below the surface to insure against the danger of just such an accident, M'koko seemingly made no effort to save himself. In a notable display of care as well as courage, George Scott, a keeper of birds who happened to be walking nearby, jumped in, and though hardly able to swim himself, tugged the huge ape to dry ground, where another keeper tried vainly to resuscitate him. After other zoos—Frankfurt among them —had similar tragedies with gorillas, water barriers for their enclosures were either replaced by thick glass or constructed with more of a thought for the great ape's total helplessness in water.

Another grave error in design occurred a while back in a German zoo that had gone to considerable lengths to accommodate its polar bears with special caves to which a prospective mother could return for cubbing. Soon after the young were produced, both cubs and mother were found dead of hyperthermia in a retreat that had become too hot and humid for them.

In caring for zoo inhabitants, smaller details than these take on importance. An aviary, for instance, must provide its birds with a variety of perches, such as birds would find in nature, to keep beaks and claws in trim. But these perches must be kept free of irregularities that might injure a bird's feet, raptors being especially subject to serious infections. A particularly imaginative touch is the use of bright red dowels in the food dishes of certain young precocial birds. Pecking away at the attractive object, the artificially hatched chicks inadvertently take in a bit of the food, enjoy it, and so learn to eat without having to be fed by hand.

A certain toll of injuries inflicted by dominant male animals on challengers within the herd, and vice versa—although nothing like as many as are inflicted in the wild—has to be expected among the hoofed creatures of a zoo. Unable to afford the luxury of many such casualties, keepers are alert for contests in the rutting season, and will generally try to forestall such engagements by judiciously removing possible contenders from a paddock before real damage has been done.

Nevertheless, such injuries do occur and have to be treated. The following passage from the periodically issued *Scientific Report* of the London Zoological Society gives some idea of the extent of intraspecies (sometimes interspecies) warfare in a major zoo. Here the damage was discovered only by post-mortem examination:

> *Injuries inflicted by other members of the group were found in an old female moose, an adult female brindled gnu, a Japanese sika deer stag, a female Père David's deer, a female Chinese water deer and a female Cape hunting dog. A Mouflon ram and a breeding male llama both died with penetrating thoracic wounds inflicted, in the case of the llama, by a nilgai bull . . .*

Warren Thomas, director of the Los Angeles Zoo, started tipping the horns of his oryx when he found that he was losing one or so a year in combat. Didn't that interfere with natural selection and perhaps prejudice the best propagation? "Not necessarily," he said. "The animal manager decides which male will sire the next generation anyhow"—a choice that is still a gamble.

> *When you don't have enough of a breeding record, you may inadvertently pick an animal that appeals to you—one with longer horns, brighter color, longer tail, or smoother coat, but that tells you little about its genetic makeup. In the wild it could be some ratty looking male, resembling the last rose of summer, one you'd knock on the head, that would be the best to sire the next generation. So in reality you're gambling and only time will tell.*

Even so, he draws the line at pulling a weak male out of the herd to be hand raised. "They have to make it on their own. If they wouldn't make it in the wild, they shouldn't be making it here."

At the zoo in Basel the vice-director, Hans Wackernagel, has some doubts whether the effort should even be made to keep delicate animals like the high-strung proboscis monkey, which requires a temperature of around 80 degrees Fahrenheit and a humidity close to 100 per cent. But, he adds—and this is what medical change and technology have achieved in the zoo—"Twenty years ago I'd have said, don't keep penguins. Now we have climatized cages, get fresh fish from the sea, and can keep them as easily as any other animal."

An essential improvement, says Dr. Peter Weilenmann, his counterpart in Zürich, is the use of tranquilizing drugs, in connection with which the dart gun has "solved a lot of our problems. Usually, but not always, darting is better than the old squeeze cage, into which the animal had to be lured and which was then compressed until the patient could not move in any direction. The anesthetic was then administered, but if the poor animal was not in a state of shock by that time, it was badly frightened and very likely to hold the whole procedure permanently against all those associated with it.

Ideally, darting does away with that terrible tension, but it presents such dangers and difficulties of its own that zoo people use it sparingly. Not only has each species its own resistance to a particular tranquilizing drug—a 200-pound young bear at the National Zoo was found to require six times the dosage of M-99 needed to knock out a 900-pound polar bear —but the general health and metabolism of the individual animal are factors as well. "Any time you introduce a drug into an animal you have to be prepared to pay the consequences," I was told by Ed Alonso at the Los Angeles Zoo. "You're taking it to death's door and hoping to bring it back again, and that can mean walking a very thin line."

A constant exchange of information from zoo to zoo is beginning to provide an index of experience regarding proper dosages for different animals and the time required for their effect. Even so, the body of knowledge needed is still only meagerly developed. Where possible, immobilizing drugs are administered in food—a procedure allowing for a more gradual approach—or, as Alonso much prefers in the case of a minor medical job—simply by getting a hand-grab on the animal and bringing it down. Obviously this technique is limited to certain species, mostly antelope, and is hardly recommended for gorillas or Kodiak bears. Once down, the animal is quickly blindfolded, since darkness has a calming influence, and given a local anesthetic for treating superficial wounds, bites, cuts, and the like. Tranquilizers are resorted to only if more serious work has to be done.*

---

*Reluctance to use a dart gun on a powerful carnivore may not unreasonably spring from the fact that one can't be certain whether it has taken full effect. Sheldon Campbell in his *Lifeboats to Ararat* recalls a San Diego veterinary who shot a full-grown Bengal tiger

What leading zoos do today with the remains of their deceased boarders is almost as important as the studies they make of them alive. The Penrose Institute at the Philadelphia Zoo, which pioneered in wild animal autopsies, can show you in its files sample tissue from the 25,000 animals that have died there since 1901. Preserved in small parafin squares, these exhibits have provided invaluable data on wild animal disorders. Just as impressive are the records, autopsy rooms, and laboratories of the San Diego Zoo, as well as its sick bay, through which I was guided by the zoo pathologist. We inspected a marabou stork with a gangrenous toe, a beak-clacking great horned owl, and an indisposed leopard that managed to direct a stream at the pathologist, who scientifically interpreted the gesture for me as a friendly overture.

As a consequence of improved medical attention and diet—in addition, of course, to security from attack and well-planned housing—animals in the better zoos outlive by far their counterparts in nature. To drive home the point, Dr. Eisenberg at the National Zoo drew a numerical picture for me of the comparative fate of a hypothetical group of toque macaques in the wilds of their native Sri Lanka and a real such group in the zoo. Imagine fifty of these monkeys, equally divided as to sex, born in the wild on the same day. By the end of the fourth year you would be down to ten males and five females—"the young die like flies in the wild." By the eighth year, three females and one male would be left. These, eight per cent of the original group, would then be good for another decade. In the zoo, by contrast, forty monkeys were still going strong after the fourth year and thirty-five after eight years—that is, 70 per cent of the starting group.

Obviously, the longer animals live, the better the chance of progeny (with the exception of some few species that reproduce very early and then lose all sexual drive). Just as obviously, the more effectively a threatened species propagates in captivity, the more we must rely on zoos to keep it going, rather than risk its survival on the high mortality and declining birth rate of an ever-shrinking wild.

with a drug intended to allow a leisurely look at his swollen jaw. The tiger, which had appeared to be asleep, had just enough consciousness left to get up on his feet and claw the doctor's shirt, pants, and skin before the drug happily caught up with him.

Zoos and Zoo-Goers

Zoo: An excellent place to study the habits of human beings.
—EVAN ESAR

If the breeding of threatened species were the whole purpose of zoos, they would do well to say so and then proceed to close their gates to the public. Thus freed from the expense and concern of having to serve the needs and pleasures of zoo-goers, they could devote themselves solely to the arts of wild animal husbandry. There are a few such establishments now, and the hope is that there will be more. But to reduce all zoos to this single role, vital as it is, would be to cancel out an equally vital purpose, indeed a major reason for their existence.

That purpose is to stimulate the feeling for wild animals, which in an increasingly urban society grows fainter by the decade. To man in an urban setting, the zoo represents a link with the forest world of his origins. Without an occasional eye-to-eye contact with wolves, bears, tigers, and such, he will see their pictures in an encyclopedia with no more emotion than he gets from those of an archeopteryx or mastodon; they will mean to him no more and no less than those vanished creatures, and when these too disappear, he will only wonder that they had so recently existed.

So it is that if the zoo is to conserve wildlife, it must be more than a survival center; it must be a school and observatory. Not least, it must be meeting ground between the human animal and those wild creatures that may be regarded as envoys from the disappearing biotopes of his own planet. Let us say they are there, in the zoo, to invite our interest and good will, without which they are doomed to a short future and we to an impoverished one. "I am most concerned with zoos as education," said Jaren Horsley, a general curator at the National Zoo, seeing zoos principally as a "form of theater that can enable people to feel a strong affinity to wildlife."

Coherently arranged and fortified by bright and instructive signs and labels, a zoo's exhibits can go far toward achieving that purpose even if it is limited in size and budget. Having either to confine itself to a few species or do an inferior job, the small zoo will ideally pick them in the first place for clear and appropriate reasons. It might content itself, for example, with showing city children the small animals of their region which, in the very nature of urban and suburban sprawl, they would otherwise rarely if ever see. Or it might demonstrate the relationship of particular animals to their habitat, showing prairie dogs and their burrows or beavers at work on a dam. Limiting itself, in an extreme case, to a single class of animal, it might have several good-sized aviaries with enough color, variety, and interesting information to allow visitors to really enjoy birds for the day instead of hurrying past them on their way to the apes and elephants. Or it might even feature insects, such as bees or ants, under magnification as they go about the affairs of their complex communities.

Major zoos with ample space obviously have wider choices in what they offer. Traditional ones, now mostly outdated, distribute their charges simply by animal type: big cats in one building, with outside cages attached; primates in the monkey house; elephants, rhinos, and hippos in a concrete compound of their own; and so forth. Some of that rigidity has happily disappeared with the spread of the open zoo, with its islands, plains, woods, and other appropriate habitats, each surrounded by a moat, a steep wall, or some other natural separator. Such a zoo may choose to divide its exhibits by continent or some other geographic area. Or it might choose to arrange its animals according to behavior, concentrating in one locale, for instance, those that sleep by day and come alive by night. Many a visitor to the old-fashioned zoo was frustrated to read the label and look in vain for an animal happily retired for the day in a hollowed-out log. Thanks to red light and other special effects that reverse day and night, such creatures—lorises, bats, fennec foxes, civets, and many others—can now be seen in a number of zoos going about their nocturnal business when elsewhere in the park all is bright sunshine.*

Fine examples of education through imaginative zoo layout, combining geography and natural animal relationships, are the predator–prey exhibits of the Milwaukee Zoo. Here are lions on a formation that overlooks another area but is separated from it by a moat 22 feet across and 15 feet deep. At the lower level are the lion's natural prey—antelopes and zebras—all aware that, however vulnerable they may appear to the zoo-goer, who barely sees the moat from the front of the exhibit, they are safe from the lions; and the lions are too well-fed, if not too well-bred, to be frustrated by the arrangement. Similarly, the South American display features jaguars above and behind their usual victims, such as tapirs and capybaras; and in the Asian exhibit Bengal tigers seem all but on top of herds of blackbuck, axis deer, and Indian waterfowl.**

The imagination, money, and effort that go into displays of this sort are spent far less for the satisfaction of the animals on view than for the education and pleasure of their viewers. It is for them that habitats are simulated, that the surrounding park is landscaped with the botanical displays that are so notable at Chester (England), Rotterdam, and San Diego—that zoo design and architecture, in short, have come to assume an importance they never enjoyed in the old days. As for the animals, says David Hancocks, director of the Seattle Zoo, who is probably the only architect to serve in that capacity, "many of them could be bred just as well in stables and barns, with a few screens for hiding."

Yet no zoo official belittles the investment in either esthetics or education. Their facilities, they know, are supported by the public—directly by admissions, indirectly through taxes—and only the pleasure and interest that the public takes in them can assure the financing needed to conserve wild animals.

In my wanderings through more than two dozen zoos on both sides of the Atlantic, I saw a score of exhibits that remain vividly with me as splendid examples of this blending of esthetic pleasure and exciting education in the ways of the wild. Three of them are in the Bronx Zoo alone:

*A particularly sound innovation inasmuch as 60 per cent of all land vertebrates are creatures of the night.
**Such optical illusions are worked into the design of many zoos, but no book on the subject can fail to note that all of them owe their inspiration to the famous Hagenbeck family's Stellingen Gardens in Hamburg, the first zoo in the world to display animals without visible barriers. In the art of exhibiting wildlife it was a pioneer, and I greatly regret the circumstances that made it impossible for me to pay it a visit rather than merely my respects.

The World of Birds. For educational skill, showmanship, and sheer delight, few exhibits in the zoo world can match that park's cylindrical cluster of brilliantly replicated bird habitats. For the viewer, from a high walk bridging a South American rain forest, life in the treetops is especially stunning during the intermittently staged thunderstorms that send the quetzals, tanagers, and toucans rushing to seek shelter in the foliage. They are free to fly where they will in an enclosure that needs no internal barriers because the simulated habitats are so much more enticing than the walkways for the observers.

Gibbon Island. A few feet offshore, in a small stream thick with waterbirds, are two tiny islands on which at almost any time of the day you can see the world's most agile acrobats. Along a rope linking the islands, as well as among the branches of the tree, a small group of gibbons do every trick known to trapeze artists and a great many more as arm over arm, they swing, or "brachiate," sometimes seeming to fly as much as seven or eight feet between handholds. Occasionally one will dip its fingers into the water or reach out in a futile grab at a passing duck.

Wild Asia. From a low, slow-moving, frequently stopping monorail, the visitor can look out on a succession of well-planned reproductions of life along the Irrawaddy, the Brahmaputra, and points east. A female tiger and her cubs lap water from a pond in one section; in another, young specimens of the rare Indian rhinoceros come galloping toward the monorail, as curious evidently as the passengers. Further along, Formosan sika deer may be seen browsing, as they no longer do in their vanished native habitat.

Happily the Bronx Zoo has no monopoly on memorable scenes of this sort. I recall with delight the king penguins strolling along a path at the charming zoo in Basel, their keeper following at a distance like a teacher with an eye on her unusually well-behaved pupils. Among the scores of other remarkable displays that stay in the mind of a zoo buff, here are a few:

The Eleanor S. Gray Memorial Hummingbird exhibit in Philadelphia, where one discovers not only the structural beauty of these gem-like creatures as they dart about in the free-flight aviary, but their ferocious sense of territory (a population of twenty-four, I was told, had reduced itself in battle to half of that in a matter of six weeks—after which there was ample room).

The tigers at the Busch Gardens in Tampa, viewed from a slight elevation as they play on a wooded promontory, splashing in the surrounding water or scampering over the ground in what appears to be an existence far more relaxed than that of Philadelphia's hummingbirds.

Penguins at the Frankfurt "Exotarium" in a perfect replica of the Antarctic landscape—a glass-enclosed exhibit enriched with ozone, furnished with natural ice, and kept at a constant temperature of 8 degrees Celsius (47 degrees Fahrenheit).

The Toronto Zoo's underwater show of polar bears, which may be watched—for hours—through an outdoor glass wall as the huge creatures glide to within inches of a viewer before veering off.

The unwalled compound at Whipsnade, with its long Bedfordshire vistas, in which yaks, onagers, Przewalski horses, and a few other species enjoy a paddock big enough to accommodate the entire London Zoo.

The richer they are, the more such displays pique the viewer's wish to know more. It is here that the graphics of a zoo come into play. Like the exhibits themselves, they have changed considerably since the days when zoo-goers stolidly encountered such sparse enlightenment as "Lion *(Panthera leo)*. Habitat, Central and East Africa," or "Polar Bear *(Ursus maritimus)*. Habitat, Arctic regions"—possibly extended by the additional datum that "Whitey," the cub, was born on the premises.

The thought then was that since people came to the zoo solely to be entertained, why burden them with unsolicited information? As zoo directors became more imaginative in what they showed and how they showed it, however, the temptation must have grown to generate a little more appreciation of their efforts. No use having an electric eel if you didn't really let the public in on its precise capacity to shock; a shame not to let visitors know exactly how much food Jumbo consumes in a day.

As the feeling grew that a few interesting insights need not be incompatible with having a good time, graphics gradually improved, especially where they were taken over by newly established education departments. Today in the better zoos, these signs have reached a degree of sophistication that both adds enjoyment and promotes the idea of conservation. The simplest and most obvious sign will now usually include such information as the geographical range of the animal, the nature of its habitat (jungle, savanna, swamp, rain forest, mountains, or desert); its diet in the wild and in the zoo; and something of its social habits—whether it is a group animal, a loner, or possibly sometimes one, sometimes the other. Any zoo worth its salt will also now indicate by a symbol used throughout the park those animals that are endangered.

Beyond these essentials, zoos are increasingly trying through graphics to give the visitor something of a feel for animal lore, scientific reality, and the environmental state of the world. In the first of these categories, for example, the Los Angeles Zoo tells viewers of its maned wolf enclosure: "It has been reported that some South Americans make a tea of bone shavings from this animal. This concoction given to pregnant women is said to ease delivery." The observer may smile patronizingly, but he will appreciate a little better what the maned wolf has been up against. A typically enlightening sign in the Bronx Zoo is the one for the giraffe enclosure, depicting a skeleton of that animal's neck alongside that of a bird's neck under the heading "Basic Designs." From the brief accompanying text, one learns that "all mammals are built on the same pattern . . . The long neck of a giraffe contains only 7 bones, the same number that we have," while "a typical bird, the pigeon, has 14, making its neck more flexible than that of a mammal."

Allowing for varying degrees of curiosity in its guests, Chicago's Lincoln Park Zoo has installed revolving graphics for some of its exhibits which, starting with a few simple facts, yield increasingly complicated information to anyone interested enough to keep turning. Preparations were also being made, when I was there, for putting up a colorful wall panel showing, step by step, the evolution of reptiles, the kind of display once more likely to be seen in a museum or classroom than in a zoo.

Some zoos have become unabashed preachers of conservation. A graphic in Los Angeles serves the cause by pointing out how a particular animal has benefited the ecology: "Dingoes have done Australia a great service by controlling the number of rabbits." One at the Bronx's jaguar enclosure lashes out at the needless threat to a whole family of creatures. "There is more than one way to get a cat's skin," it reads, adding that in 1968–69 alone "306,035 big cats were shot, snared, leg-trapped and poisoned for the U. S. fashion market—3,168 cheetahs, 17,490 leopards, 23,347 jaguars, 262,030 ocelots," concluding with the pleasantly unscientific commentary, "DISGUSTING!"

Other signs, also in the Bronx Zoo, concern the environment in general:

Each year in the United States 1,200 square miles of animal habitat are replaced with asphalt and concrete . . . Each year Americans contribute 50 billion cans, 28 billion bottles, and 58 million tons of paper to rapidly growing dumps. These dumps are located in areas which once were rich in wildlife.

Do zoo-goers read these injunctions—or any graphics, for that matter? David Hancocks, citing a Brookfield study that showed the average time spent on a graphic to be four seconds, thinks the question needs a lot more investigation. He looks with disfavor on signs that appear to be "a page taken from a textbook and blown up." But an elaborate study made by Dr. Neil H. Cheek, Jr., supported by grants from the federal government and the New York Zoological Society, showed that of the 1,251 people he interviewed, 69.7 per cent read zoo labels, and roughly half of these listed reading them as an activity they particularly enjoyed. It was also demonstrated at Brookfield, moreover, that as with much else, the public responds to quality in graphics. Good signs (presumably lively and interesting ones) were found to elicit twice the attention span of poor ones. What seems beyond dispute, judging from the burgeoning of their education departments, is that zoos now see an obligation to enlighten as well as to entertain their paying guests.

Yet zoo educators have to keep clearly before their eyes an objective that differs essentially from that of the classroom. The facts of natural history are important not because they satisfy some academic requirement but rather for the feeling they convey that wild animals are appealing and important in their own right *and* for their vital relationship to human life. Ideally, zoo-goers will get to know the animals by sight, hearing, smell, and in some circumstances by touch. In the process they will welcome and absorb as much fascinating objective information about them as their increasing involvement demands. From these experiences, the hope is, will come the future champions of conservation—the goal that educationally distinguishes zoo teaching from the work of schools.

Troops of touring children have long been a common sight in the parks of the world, but such tours are now merely the beginning of a zoo's education program, which provides continuity rather than an occasional outing for teacher and pupils, with entire courses on distinctly different levels for children, adolescents, and adults. The following entries in the Bronx Zoo's brochure for adults read like those in a college catalog:

Primates and the Evolution of Man. What is the relationship between a galago and baboon, a gorilla and man? This course will help you understand the ideas that biologists believe may explain the evolution of primates, our closest relatives, and higher verte-

brates. Controversial and complex issues such as speciation and natural selection will be examined in depth, but presented in a manner easily understood by a person who is not a biologist.

Vanishing Habitats. . . . In this course you will learn whether desert animals can manage to coexist with nomadic tribes in their harsh environments, how rain forests are being cleared for timber, endangering the survival of many spectacular creatures . . . Each of these self-contained units will concentrate on a different habitat and its inhabitants. Live animals, films and exhibit observations will supplement each unit.

For the lucky children whose parents will put up the fee, there is a five-day Zoo Camp, each day devoted to the wildlife of a different continent. The luckiest ones, for an additional fee, can enjoy an overnight camp complete with "a cook-out, a singalong, and, best of all, a flashlight safari through the Zoo" under the magic cover of darkness. Similarly, the Philadelphia Zoo features summer courses and workshops ranging from a two-hour session about pets, for four-year-olds, to a two-week program on animal behavior for youngsters from 11 to 14. Advanced pupils are taught to observe animal life at the zoo and report their findings, with films and discussions of Konrad Lorenz and Jane Goodall to inspire and set the pace. From the ranks of its regular teachers, the city selects—and pays—eight "museum teachers" for permanent duty at its cultural shrines, which include the zoo in Fairmount Park along with Independence Hall, the Franklin Institute, and the Art Museum.

Even though these great facilities are exceptional there is hardly an animal collection now so poor or laggard that it does not have some sort of educational program, or at least a pretense of one. Many municipally financed zoos find it hard to squeeze the extra dollars out of city or county politicians, for whom too often schools are work and zoos are fun, and never the twain shall meet in the local budget. The education office of the Milwaukee Zoo (a county institution) had to start out in 1977 on an appropriation of only $2,000 and in 1980 was doing as well as could be expected on the still dismaying budget of $15,000. Until a few years ago, Seattle's zoo had no education personnel at all, and when its director, David Hancocks, finally succeeded in getting the green light to appoint a staff of two, he could pry out of the City Council no more than $200 for their supplies. Allowing a bit for hyperbole, he informed that body that if he could spend no more than that on his education program, everything else spent on the zoo would be a waste.

A more fortunate, and less typical, small-city zoo is Roger Williams Park in Providence, Rhode Island, which committed energy and talent to its education program while most zoos of comparable resources were investing little more than lip service to that cause. As a zoo introduction for the young, its booklet, "Zoo Animals and You," is a model of clarity, conciseness, and pleasing design. Like many zoos now, it has issued to visiting classes of school children a stream of quizzes, work sheets, and puzzles directed to various age groups.

On another level, inevitably, are the guidebooks of the great zoos, far richer productions addressed to more sophisticated readers. Outstanding among these guides are National's *Zoo Book*, filled with historical bits and good pictures, the *Philadelphia Zoo Animal Book*, and San Diego's *Wild World of Animals*, which besides telling all about the Zoo and Wild Animal Park, is a fine introduction to the excitements of natural history, with simple and well-illustrated sections on ecology, adaptation, and man's place in nature.

In the best of educational programs the zoo visitor's eye would be supported by his ear. "Ideally," Hancocks suggests, "you would have someone standing next to you, interpreting what you see, so you would get visual and aural impressions at the same time; that's how television works." Obviously this is not possible, but some zoos approach it by making it easy for visitors to question zoo personnel, including volunteer docents, placed on duty at strategic points. Other zoos, such as the one at Frankfurt, which boasts of being the state of Hesse's "greatest educational facility" (160,000 school children a year put the finishing touches on their biology course there) have installed at various points equipment that for a small fee provides recorded commentary on the exhibits. An article in the *International Zoo Yearbook* for 1975 describes ways in which the Honolulu Zoo provides for viewer participation in one of its demonstrations of animal behavior: A nylon rope, passed through a series of guideposts with a yard or so of slack, allows a visitor to engage in a tug of war with a gorilla. The contact clearly gives pleasure to both human and ape, besides displaying the animal's awesome body in action. The gorilla's muscular triumphs must refresh its ego even while they relieve its boredom.

Honolulu also offers its small-fry visitors a monkeys' island of their own, complete with climbing and swinging apparatus, as well as a tower to enable visitors to view giraffes at eye level, and thus to see the world from their vantage point. With this same approach to natural history through demonstrations, the Bronx Zoo plans to activate its wolf wood once or twice a day. By blowing in turn on differently pitched whistles, to which particular animals have been trained to respond, the keeper will summon wolves to the front of the enclosure. If it works out, the demonstration will serve at the same time to bring the wolves out of hiding—as some are sure to be doing at any given time—and to show the acuteness and discrimination of their hearing.

The lives of zoo personnel would be less exacting than they often are if a great many visitors did not have to be taken care of on a somewhat lower level than we have been considering. In a relaxed moment, a zoo official is likely to drop a hint or two about the burden of guests for whom ecological education would have to start a long way back. An assistant at the Marwell Zoo, writing in its publication that the zoo's "education service is a vital one," conceded nonetheless that "we have to deal with the thousands of school children who visit us each summer [and] sometimes we would like to deal with them by throttling them."

The comment was uncommonly blunt, but almost any zoo keeper in the world, diverted from the problems of creatures *expected* to be wild, will understand it. In *Zoo—Animals, People, Places*, Bernard Livingston quotes a zoo director: "A great many of us are anti-people, you know. Zoo-keepers see enough mischief in individual human beings to, rightly or wrongly, turn them off people generally. It's a fantasy of course, but I guess deep down many of us dream of a zoo without visitors."

Of more concern than school children as such, however, are the vandals, in school or out; those who do their best to upset the animals, especially those who offer them lethal foods; thieves; and those showoffs who insist on seeing how close they can come to a lion or a grizzly bear without getting mauled.

Sadists evidently know no geographic or political boundaries. A few years ago, not long after five fallow deer were beaten to death in New York's Central Park Zoo, a family of grey kangaroos were fatally knifed at the Moscow Zoo. The Soviet newspaper reporting the incident added that there had also been a number of recent thefts—including that of a penguin carried off in the subway—and a foiled attempt to steal twenty birds valued at $3,000.

The would-be bird thief in Moscow, it might be noted, drew a six-year sentence, rather severe compared to the penalty meted out to some Philadelphia youths who had stoned to death a half dozen flamingos in

Fairmount Park; they were ordered to take educational courses at the zoo, in the hope that they would learn to value animals. Elsewhere, I learned that four macaws had been stolen from the free-flight aviary at the Milwaukee Zoo, six eggs of the Cheer pheasant were taken at Marwell after two weeks' of incubation, and Whipsnade lost several eggs of the very rare Manchurian crane, presumably for sale to unscrupulous collectors.

Accidents to zoo-goers as a result of their own folly are relatively rare in proportion to the prevalence of that human attribute, but they do occur. Newspaper readers were horrified a few years ago to learn of the small child who was dragged between the bars of a lion's cage and killed when an unbelievably careless adult allowed her to climb over the railing at the National Zoo. Two innocent polar bears had to be shot in 1979 after a youth somehow managed to fall into their pit at the Buffalo Zoo. And a visiting Israeli was lucky to escape with his life when, ignoring all the warnings, he kept his car window open in a dangerous area of a New Jersey safari park and was severely clawed by a lion.

The commonest miscreant is of course the zoo-goer who feeds the animals, a pastime theoretically allowed almost nowhere and practiced in fact almost everywhere. Most of it is done by well-meaning but unthinking people who can see nothing wrong in getting bears to beg for candy or elephants for peanuts. A much annoyed woman who was ordered to stop feeding bears by the director, with whom I was touring the Milwaukee Zoo, irately demanded to know what harm she could possibly be doing by giving the creatures tidbits that they so obviously enjoyed. It had not occurred to her, as it has not to millions of others, to multiply her donations by the number of similarly inclined visitors. The calculation would have revealed a volume of unprescribed food far greater than the total diet so carefully figured out to keep the animal in good health.

Even a food harmless in small amounts can do great damage when consumed in quantity. Elephants have died from overdoses of sugar. What can be said, then, of the chocolate-covered razor blades, marbles, and lighted cigarettes fed to zoo animals by perverts—or the coins tossed to seals by those who manage somehow to confuse their tanks with wishing wells? In a post-mortem examination at the San Diego Zoo, the stomach of a harbor seal yielded $2.57 in pennies, nickles, and dimes, not to mention a few Mexican pesos.

Apart from the specific and obvious harm that it does, feeding by zoo-goers has other extremely negative effects. It can destroy an animal's appetite for the foods it really needs, or bring on an unhealthy obesity. It adds grossly to the littering of the park with candy wrappers, Crackerjack boxes, and the like.* And, not least, it leads to behavior problems among the recipients—jealousy, fighting, and the habit of begging, which only the uninitiated can consider natural, much less engaging.

Considering all aspects of the visitor's requirements—the need to entertain, instruct, feed, and police him—it should surprise no one that, as Dr. Reed says, "nearly 80 per cent of a zoo management's time, effort, and money go into taking care of human beings." Yet, taking the long view, few zoo directors would consider the investment unsound. Many captive animals come to relate to their human audiences and miss them when they are gone. Hediger, a firm believer in the importance of the visitor's role, told me of the boredom and passivity of many of the animals when the Zürich Zoo closed for a time because of a hoof-and-mouth disease epidemic. Most affected, naturally, were the extroverted species—the Capuchin monkeys, elephants, and big cats, for example. Interestingly however, others of the extroverted monkeys found enough company and excitement in their own highly social groups.

Going far beyond the education of its visitors, zoos are potentially a source of scientific knowledge that can affect people who never go near them. Studying animals in their native habitats undoubtedly has great advantages, but it cannot always be done. The incomparable work in the wild by the Schallers, Goodalls, Paynes, and others, has mostly focused on large animals—lions, gorillas, chimpanzees, whales—whose size and habits have made such study possible, however arduous. It has not been comparably practical to study in detail the behavior in nature of small, elusive creatures, carrying on their lives and functions in the extreme privacy of burrows, tree-hollows, and the like. In a well-ordered zoo, such animals can be observed at leisure and with perfect continuity; contraptions for looking in on them, measuring devices, and record-keeping systems can be used over an extended period. There are now more than a half million wild vertebrates in some five hundred zoos and aquariums around the world. This matchless reservoir for studies can lead to the conservation of species, perhaps to a knowledge of the best

---

*In this connection zoos everywhere might take their lead from a sign in a small, rather pathetic animal collection on the outskirts of Nairobi. I cannot recall the precise wording, but it was couched in the formal English so prevalent in what was once British Africa and read something like this: "Any Visitor Found Discarding Rubbish in the Crocodile Pit Will Be Obliged to Retrieve It."

ways to reintroduce them into restored habitats, and to all sorts of new techniques for observing their behavior, hear Conway:

> *The zoo, perhaps, is most easily utilized for comparative studies on a broad spectrum of species; its special facilities can provide an important supplement for intensive field investigations of a particular species and can suggest new approaches to old problems. The biologist familiar with marsupial reproduction finds mammals available for observation at an extraordinarily early stage. The giraffe watcher may wonder how this walking scaffold avoids a stroke when he lowers his head eighteen feet in one swift movement. . . . Navigation may be studied in experiments with birds which migrate over trackless oceans at night, echolocations with bats, and learning and transmission of song patterns in birds can compete for the investigators' attention with the problem of animal play and the evolution of facial expression. . . .*

At this stage of their history it can hardly be said that scientific research is a general concern of zoos. Extremely few have the facilities, the expert personnel, or the money to engage in major research programs, nor does the public expect it of them. On the contrary, the very mention of animals in connection with research is a red rag, arousing suspicions of vivisection and other inhumane experiments even though zoos would be the last institutions that would attempt, or could afford, any such risk to their costly collections.

Nevertheless, some of the world's greatest zoos have long distinguished themselves scientifically through close affiliation with institutions that are in every way equipped to do advanced work in the areas of wild animal biology and behavior. Since 1905 the bodies of all dead animals at the Philadelphia Zoo have gone to the Penrose Research Laboratory, which is also attached to the Medical School of the University of Pennsylvania, for detailed and microscopic study.* Until the work done in the research program at Penrose, it was not generally believed that such creatures as lions, zebras, elephants, and apes were susceptible to tuberculosis, hardening of the arteries, heart disease, cancer, and other ailments so often associated with man's perverse habits and unnatural ways of life.

---

*Experimental surgery on live animals, feeding experiments, and even the manipulation of animal groups for behavioral studies are strictly forbidden, but blood samples and other body fluids may be taken when an animal has to be restrained and treated for other medical purposes.

Alas, they are just as susceptible. In the wild, animals die at so fast a rate—from predation, failing food supply, and other ecological causes—that without the experience of captivity, their proneness to our familiar diseases would never have been known. At the San Diego Zoo's elaborately equipped laboratories it has become apparent that cancer occurs at a steady rate in aging animals, just as it does in humans. Comparative pathology is a major study at the zoo's research center, headed by Dr. Kurt Benirschke, who formerly directed the pathology department of the University of California Medical School. Students still go back and forth between the university and the zoo laboratories; the day I visited, two of them were dissecting a scimitar-horned oryx that had just died from unknown causes.

It was there that research into chromosomes showed the okapi not to be a close relative of the giraffe after all, but probably—though there is still no certainty—closer to the antelopes. Similar chromosome studies at the Rotterdam Zoo are being done on the douricouli and are expected to be highly useful in the saving of that South American night monkey. San Diego's research division also collects and stores sperm in liquid nitrogen, not only for future research but also (who knows?) for possible revival at a time when the species no longer exists. And the facility keeps a close watch on genetic diseases so that possibly affected animals will not be mated in circumstances where there is danger of harmful inbreeding.

Under the aegis of the London Zoological Society, besides its own Institute of Zoology, are the distinguished Nuffield Institute of Comparative Medicine and the Wellcome Laboratories of Comparative Physiology. The first of these is charged with maintaining the health of the zoo's resident population; Nuffield is concerned with the diseases of animals, whether wild, domestic, or human; Wellcome concentrates on the reproductive processes of mammals and on captive breeding. Lord Zuckerman, secretary of the London Zoological Society, states the case for this most comprehensive of all research centers for animal biology and medicine:

> We have to back up our efforts to breed endangered species with an expert knowledge of reproductive physiology, of animal nutrition, and of animal disease. . . . Without a good understanding of territorial behavior, unless we bear in mind some idea of the selective forces which have brought about the spread of animals across the globe, the reintroduction of animals to the wild might well prove a wasteful and cruel enterprise.

Others of the world's zoos, while not scientifically equipped on the scale of these three, fund and sponsor research programs in varying degrees—among them Antwerp, Rotterdam, East Berlin, Brookfield, Oklahoma City, Portland, and the National Zoo in Washington. Besides bringing in scientists to observe and study, they sometimes launch young zoologists on their professional lives. At the Frankfurt Zoo, Dr. Faust told of a graduate student who did his thesis on warthogs there, went on to Nairobi, and is now launched on his scientific career, and of another now a successful ethologist, who did his early work at the zoo because he needed to observe animals under controlled conditions—a practice that has become quite common.

The most far-flung efforts of all are those of a few zoos—the Bronx, National, and Frankfurt preeminent among them—to sponsor scientific studies of animals in the wild. Nothing could be more basic to either public understanding or to the effectiveness of zoos themselves than to learn as much as can be found out about the behavior of animals in their native habitats—their diets, sex lives, social organization, and living habits. Beyond these purposes, exploration is vital if we are ever to know how to reproduce a habitat.

In the past few years of the New York Zoological Society, funds have gone into projects ranging from a protection program for logger-head turtles on Georgia's Ossabaw Island to a survey of the pileated gibbon in Thailand—not to mention the projects carried on by its own staff members. Roger Payne, who put the voice of the humpback whale on phonograph records, has been surveying the situation of whales from the North Pacific to Patagonia. Thomas Struhsaker has been studying primates and rain-forest ecology in Uganda. David Western's work on a water diversion scheme and ecological monitoring system has already gone far to save Kenya's Amboseli National Park. And George Schaller, who in previous projects for the zoo added greatly to man's knowledge of lions and snow leopards, later shifted his focus to the jaguar and other mammals of the Brazilian jungle and most recently to China's giant panda.

Invaluable as all these projects are, however, it is not for scientific research and operations that in the end zoos will be judged, but rather for the degree to which they bring wild and human species together to their mutual advantage.

Charges and Rebuttals

Nature has been described as a treasure house of knowledge, and zoos are the caretakers of a considerable part of it, but only too often they are unaware of their responsibilities as caretakers, or even that they are caretakers at all.

—CAROLINE JARVIS (LADY MEDWAY)

For the most savage, sprightly, and happily outmoded attack on zoos, the reader is referred to H. L. Mencken, the onetime Sage of Baltimore. Asking himself, some sixty years ago, to what end taxpayers bore the burden of a zoo's maintenance, that hard-shelled critic concluded:

> *To the end, principally, that a horde of superintendents and keepers may be kept in easy jobs. To the end, secondarily, that the least intelligent minority of the population may have an idiotic show to gape at on Sunday afternoons, and that the young of the species may be instructed in the methods of amour prevailing among chimpanzees and become privy to the technic employed by jaguars, hyenas and polar bears in ridding themselves of lice. . . .**

The passage is quoted here to suggest not so much the extent to which anti-zoo ferocity can go as the elevation in status that zoos have

---

*Reprinted in *Chrestomathy,* by H. L. Mencken, from a column in the New York *Evening Mail,* February 2, 1918.

enjoyed since Mencken made his estimate. For few opponents today would take quite that ferocious a view. Much less would they accept his observation concerning zoo education—that a student of the giraffe would have to conclude from observing zoo specimens that the creature is "a sedentary, melancholy beast, standing immovable for hours at a time and employing an Italian to feed it hay and cabbages."

Even among animal welfare organizations, still considerably more critical than supportive of zoos, the traditional hostility is no longer unanimous. Besides spearheading regulatory reforms, the Humane Society of the United States (which is largely to say, Sue Pressman, who led the Society into this particular area) has made a start in classifying zoos by merit—an important acknowledgment in itself that, like other institutions, zoos can be good or bad. Vigorously campaigning to improve those capable of improvement, she has gone far to expose the kind of collections that Desmond Morris, a zoo champion as well as a severe critic, has called "scruffy little animal slums."

Over all, Mrs. Pressman determined to her satisfaction in 1973 that some 25 per cent of American zoos (many of which would not be accredited by their professional organization) ranged downward from being seriously in need of reform to gross violation of even the mild requirements of the law. But Class Two, which included the overwhelming majority of zoos, were mixes of good and not so good, establishments deserving of help rather than padlocking. Their staffs were trying hard to improve conditions and seemed to know how to go about it; their problem was that they found themselves year after year unable to wrest the essential wherewithal from indifferent city fathers and inert bureaucrats.

Though far from lacking critics in the zoo world, Mrs. Pressman has succeeded in stirring up the kind of local public support needed for action in such cities as Detroit, Tucson, Tulsa, and perhaps most notably, Los Angeles. There Mayor Bradley thanked the H.S.U.S. for its services, installed a new and highly competent director, and gave him the powers he needed to bring the zoo into Class One in very short order.

The World Federation for the Protection of Animals, based in Zürich, has similarly refrained from all-out attacks on zoos as such. With a strong lead from its administrator, Karl Frucht, it has instead put out highly useful publications on the subject and arranged congresses in which the problems of zoos, among other issues, have been helpfully aired.

Sometimes a zoo is so poorly managed, whether from lack of money or bureaucratic inertia, that the keepers themselves rebel. Such revolts were a major factor in Los Angeles, Kansas City, San Francisco, and other cities, and at least a minor one in New York City's small municipal zoos. The Prospect Park Zoo in Brooklyn a few years ago witnessed the sit-in of a young keeper who locked himself in the monkey house as a protest and informed the crowd through a window about what he considered the needless deaths of zoo animals. "We shouldn't take animals out of the wild," he shouted, "and condemn them to concrete cages in Brooklyn."

The sad story of New York's municipal zoos—in Central Park and in Queens, as well as Brooklyn—is one of constricted space, poor funding, low-paid staffs, and wholly inappropriate collections. It should become less sad now that arrangements are being made for the New York Zoological Society to take over the management of all three. At this writing, agreement has still to win final approval from the city and state governments, but the chances are good that at long last these depressing Class Three collections may be turned into specialized little zoos that will do credit to the city.

According to the tentative plan, Prospect Park would be turned into an innovative children's zoo, the greatest, it is intended, in the world; Queens would remain a regional collection of North American fauna, greatly upgraded; Central Park would be freed from the frustration of trying to care for large animals which it has neither the land nor the personnel to accommodate. Big cats, elephants, gorillas, and hoofed stock would give way to educational exhibits concentrating on social relationships—predators in relation to prey, prey in relation to their defenses against predators. With the number of species reduced, space would be available for displays of how certain carefully selected animals live in nature.

On the subject of animal collections that have brought no credit to the concept of the zoo, a word should be said about safari parks. For a time these offered themselves as a contrast to caged collections, and they are. But the contrast is not generally to their advantage.

Looked at superficially, safari parks might seem a good innovation, with admirable space, the possibility of natural settings, and some of the excitement of animals at large. But in fact they have contributed nothing to science, muddled whatever educational power they might have had, and in the end subjected animals to worse treatment than they would endure in the most commonplace zoo. Describing them as "a rather unpleasant fungus" in the zoo world, Gerald Durrell declared, much as

Hediger and others have done: "An animal can be just as unhappy, just as ill-treated, in a vast area as in a small one, but the rolling vistas and the ancient trees [of safari parks] obliterate criticism, for this is the only thing that these critics think the animals want."

What the animals want and do not get in the highly commercial safari parks is, first, the security of territories that are really theirs, which they can mark for all their various purposes and where they can establish all those essentially repetitive routines that are so important a part of their lives. Far from enjoying any such security, they are notoriously harried—in some cases by keepers armed with poles—to points along the road where they can be seen by the motorized patrons who have paid for the view. A Miami *Herald* investigation of the Lion Country Safari, in Florida, found in 1972 that animals were being locked up at night in spaces that would not be tolerated for a moment if they were on view: "As many as twenty-six lions have been crowded into one 28-by-30 foot enclosure."

What is true at night is doubly true in the long winter months when safari parks, at least those in northern climates, are a dead loss. I was told by Dr. Faust about one in Germany inside whose appalling winter facilities over thirty tigers were lost in a single year. In at least one case, the commercial failure of a safari park left the unwanted animals to starve and die. Since the decisive element in any such venture is necessarily profit, there can be no sensible expectation that anything will be given priority over the one attraction safari parks can offer—the thrill of seeing wild animals outside of cages and within a few feet of viewers huddled safely in their cars. It does not matter much to these establishments that animals whose native habitats are Indonesian, Alaskan, and South African are all to be viewed together against an undisguised and totally irrelevant background, of, say, New Jersey or South Carolina. Perhaps the brightest spot in connection with this wholly commercial exploitation of wildlife is the prospective effect of a declining supply of gasoline.

The most emotional point made against zoos today, as well as the most general, is that they are prisons—the best of them somewhat more enlightened than others but prisons all the same. The argument presupposes that freedom from captivity *as such*—overlaid as the concept is with psychological and social overtones in human society—is relevant in the life of animals other than man. True, the *conditions* of confinement are vitally important. But there is no evidence whatever that, given everything it needs—food, security, appropriate physical, behavioral, and social conditions in its own territory—a lion in a zoo will pine for the ancestral spaces of the Serengeti or a crocodile for the waters of the Nile. Rather, it is the argument of Hediger's sympathetic but thoroughly unsentimental *Wild Animals in Captivity*, it is "the free animal [that] does not live in freedom." As he points out concerning those supposedly freest of all creatures, the far-ranging migratory birds:

> *A set timetable drives them from one end of their territory to the other, in strict obedience to laws. The seasonal movements of migrant birds should not be thought of as pleasure trips; these birds are in fact compelled to go on their exhausting migrations by a fixed rhythmic cycle and many fall a victim to the dangers and hardships . . .*

As concerned as anyone in the world with the real requirements of zoo animals rather than with some abstract freedom, he lays down the essential condition for a defensible captivity:

> *If all the needs of an animal are adequately met, the zoo offers its inhabitant a man-made, miniature territory with all the properties of a natural one. The animal will then consider the territory its own: it marks and defends the area and does not feel imprisoned.*

Before ever getting to the relevant question of whether these needs really *are* met in captivity, some critics question the merit in saving a species at all if salvation lies in a zoo. Theirs is the "Let's-not-play-God" approach. If nature in her wisdom, they say, has decided that the time has come for a species to disappear, as thousands of species have disappeared before it, then let man accept that fact of life rather than arrogantly venture to arrest the process.

There might be something in this philosophical surrender to nature were it not for the fact that as a rule nature has had nothing to do with the case: on the contrary, it is man-playing-God who has brought many animals to the brink of extinction in the first place. It was not the evolutionary process that decreed an end to the orangutan, but the lumber companies that cut the forests it lived in; it is not nature that has all but wiped out the rhinoceros but an irrational human predilection for its horn. Everywhere in the world, human beings have uprooted forests, drained marshes, polluted waters, and even more directly killed wild creatures by the billion. Why should he be accused of playing God only now when he is attempting to save, by whatever means possible, the last remnants of what he has so nearly destroyed?

It can be argued, of course, that the human animal is itself a participant in the natural order of things, and that its predatory behavior is therefore part of the evolutionary process. Why round up the few remaining California condors for captive breeding, as zoologists and ornithologists are desperately trying to do, instead of quietly acknowledging that as a species their time has come? Those who oppose the captive rearing of California condors argue further that a few generations of zoo life will in any event change the proud creature into a pathetic domestic bird.* The same argument has been made concerning the whooping crane, with the prediction that in captivity it would be no more than "a big white chicken."

No such thing has so far happened to the whooping crane, although a number of zoologists and conservationists do concede that captivity may in time produce both behavioral and physiological changes. When I asked Warren Thomas at the Los Angeles Zoo about this, he said flatly that if he were confronted with two lion skulls, he could tell at once which had spent its life in a zoo and which in the wild: the latter, having had to respond to tougher environmental demands, would show the more massive bone areas needed to support a greater muscular development of the jaws. But this would be a physical, not a genetic, difference between particular specimens, for a particular cause; and Dr. Hediger suggests that such differences would be less pronounced in a modern zoo, "where feeding tends to be more in keeping with the animal's natural pattern." In any case they are less than those to be found between human individuals—between a Japanese wrestler, say, and a Harvard professor.

A factor that admittedly can produce real change in a zoo population is the phenomenon known as imprinting. Newborn individuals of certain species, if hand-reared from the first hours of life, will act to some degree on the assumption that they are of the same species as the foster parent. Considerable publicity has been given to Tex, a female whooping crane at the International Crane Foundation in Baraboo, Wisconsin, whose conditioning causes her to ignore male whooping cranes entirely,

*The National Audubon Society—unlike the Sierra Club and Friends of the Earth—is no longer among the opponents of the condor project. Convinced that there is no chance of the species' survival in the wild, it now favors temporary captivity with the hope of future reintroduction in a suitable habitat—if one can be found.

performing her ritual dances and soliciting sexual engagement solely with humans.*

Just as eating habits can be changed in zoos to minimize any physiological change, however, so imprinting can be reduced to insignificant proportions by changing the routine with very young animals. The tendency in many zoos now is to discourage hand rearing, even in cases where an infant rejected by its mother may not otherwise survive. "We want our chimps to know they are chimps, not people," says Dr. Hans Wackernagel, vice-director of the Basel Zoo. Even more stringent on the subject is Dr. Faust at the Frankfurt Zoo. "Human-imprinted animals," he said, "are biologically dead. Very seldom can you reintroduce them to their groups. Very seldom can they raise their own young later on." As for reintroduction into the wild, he added, throwing up his hands— "Hopeless."

Accepting the pessimistic judgment that some changes in captive species are probable over a period of time, we are also entitled to conclude that these are less than fundamental. As Dr. Rabb, at Brookfield, sees it, the animal's essential behavior does not change, though particular patterns may—for example, the amount of time an animal devotes to a specific function, now that it no longer needs to carry on what in the wild would be an unceasing quest for food. Dr. Reed and others make the point that after generations of domesticated forebears, a horse turned loose will very quickly revert to its species' pre-domestic ways, and "pigs will get back to the old razorback" even faster—all the way to producing young born with stripes. Cats, dogs, and goats have likewise gone feral, unchanged basically by a long species history of domestic life.

In fact, one of the major differences between domestic and zoo animals is that whereas the former are bred precisely to bring about changes for the use, pleasure, or profit of their owners, zoo animals are being more and more subjected to the kind of genetic diversity that prevents drastic shifts in the basic character of a species. Some who denounce even the most undetectable changes in a zoo creature ironically have nothing to say about developing dogs to the point where it is hard to believe that any two breeds—look at the Chihuahua and the

*The staff at Baraboo, eager to increase the crane population, takes advantage of the bird's receptivity—stimulated by the dance routine's effect on her reproductive cycle—to perform artificial inseminations. At the time of this writing, according to the New York *Times* reporter Bayard Webster, she has produced two fertile eggs. In one case the shell was too soft to hatch; the other produced an embryo that died before hatching. But the prognosis is hopeful, so on with the dance.

Great Dane—are really the same species. There are also, as Dr. Rabb
pointed out, the physical changes that man and his chemicals are constantly bringing about in insects, some of which have already evolved
a distinct immunity to pesticides.

Indeed, animals are still undergoing evolutionary change in the
wild, as an adaptative response to a changing environment. Why should
the same slight ecological variations as those that occurred in the beaks
of the finches of the Galápagos, giving Darwin a clue concerning the
evolution of species, not be acceptable in zoo animals? Those that adapt
best to zoo life will change least—which is only another way of saying
that those zoos that most faithfully replicate their animals' habitat, diet,
and social existence will produce the least drastic change in their inhabitants over the decades ahead. Zoos may, indeed, be in a better position
to maintain fixed conditions for their populations than will be possible
in natural habitats constantly challenged by bulldozers, cattle, and other
disruptive intrusions.

Inevitably the question of change in captive animals involves two
variables: the degree to which a zoo tries to minimize such change and
the likelihood of a species being doomed to extinction if no attempt is
made to rescue it through captive breeding. Is it better to have slightly
modified tigers in the world, or to have no tigers at all? For those who
are ready to sacrifice the good to an impossible perfect, the answer will
be obvious: no compromise. The rest of us might even be ready to
contemplate a world in which animals fall into three categories—wild,
domestic, and zoo.

For all the breeding triumphs that have been recounted here, some
critics, even those impressed by the record, do not see such success as
justification in itself for zoos and animal parks. If captivity inevitably led
to the return of rare species to the wild, they indicate, they might modify
their opposition in spite of other objections, but they see no assurance
that such reintroduction can occur to any appreciable extent.

Here, alas, they have a point, for there can be no such assurance.
The experience of reintroduction so far is extremely hopeful but tentative. The American bison and the wisent are successes in this respect
only if we are willing to regard the somewhat controlled preserves that
are their present habitats as wild areas. Theoretically it should be possible to reintroduce some species to their original habitats—provided
these are still intact, but that is the hitch. All too often the habitats
themselves would have to be ecologically restored, food chains and all.

And that is rarely possible. The very social and economic pressures that destroy a biotope in the first place would somehow have to be relieved even before the fantastically difficult botanical and zoological work of restoration could be seriously undertaken.

Politics is frequently part of the equation. A few years ago plans were well along to bring Asian lions from the shrinking stock in the Gir Forest of India back to Iran, where the species once flourished. The program had been all worked out when the Islamic revolution exploded. Overnight the Shah, who had favored restoration of the habitat, himself became as endangered as the Asian lion.

Starts have been made at reintroduction of several species but it is too soon to say with what chance of success. Most notable of these is the attempt, in progress as I write, to return the Arabian oryx to the deserts of the Middle East. This cream-colored, straight-horned antelope which, seen in profile, is believed to have given rise to the myth of the unicorn, once roamed much of the Middle East. Unfortunately, it became the victim of a Bedouin admiration for its elegance and strength, not to mention the odd belief that eating its flesh was just the thing for healing bullet wounds. Hundreds of oryx were shot in the 1950s by hunters armed with automatic weapons, the last live specimen having been sighted in 1972.

Ten years before that a far-sighted expedition, acting on the urgent recommendations of Lee Talbot, the American ecologist, rounded up three of the animals and made them the nucleus of what was to become a world herd. The World Wildlife Fund, launching its Operation Oryx, helped get the animals to the Phoenix Zoo, where they were soon joined by a half dozen other specimens contributed by Saudi Arabia, Kuwait, and the London Zoo. The group did so well that in a few years offspring were sent to the San Diego Zoo, which since 1972 has bred more than fifty of the species.

The big test began in 1978–79, when groups of Arabian oryx were sent to Oman and Jordan for gradual reintroduction to the desert. The results of this necessarily slow process are still uncertain. The animal is noted for its ability to go for long periods without water, extracting liquid from plants and fruits, but whether captive-reared specimens will acquire soon enough the ability of sustaining themselves under such harsh conditions remains to be seen—as does the answer to whether governments can, with the best of intentions, control their subject "sportsmen." Not least, there is also the question of whether a numerically small group of oryx will avoid the possibly damaging effects of inbreeding, even though the experience in Phoenix and San Diego allows for hope on this score.

In the American Southwest it is not only the transplanted Arabian oryx that has done well. So have blackbuck in Arizona, and the aoudad —a North African wild sheep—has proliferated enough to become a nuisance in New Mexico. A staggering number of foreign mammals flourish on the hunting ranges of Texas—at least until they are shot. Among the exotic ungulates that have been stocked for hunting on these dubious preserves are kudu, ibex, springbok, nilgai, and impala, along with wild boar, tahr, and mouflon sheep.

All of these latter have been introduced into the habitat, however, not *re*introduced. The distinction is important. Wherever foreign species are imported and allowed to run wild—or even set free on preserves so huge that it amounts to the same thing—there are possible consequences of an ugly sort for the native fauna. Walking catfish brought into Florida have become a severe pest, the mongoose is a problem in the West Indies and Hawaii, and everyone knows what the imported rabbit did to Australian farms. The most flagrant case of all, perhaps, is the effect of the nineteenth-century New Zealanders' nostalgia for the song-birds of England. Imported by the hundreds, over the decades, the English species made all but fatal inroads into the native bird populations of New Zealand. Eventually some of the English sparrows from that country were shipped to Hawaii, where in spite of bad experiences with imports on the wing, some people thought they would be nice to have.

With this point in mind, naturalists who are now working on the saving of the peregrine falcon, whooping crane, bald eagle, and California condor hope to restore them to their own original habitats. Here again it is still too early to predict success, although the hatching of two bald eaglets in the summer of 1980, from captive birds released in upper New York State, is at least promising.

Generally speaking, the zoo officials I met, both European and American, were merely cautious about reintroduction, if not altogether negative. At Whipsnade, V. J. A. Manton called the idea "a pious hope." John Eisenberg at the National Zoo thought it "impossible" on any appreciable scale. Frank Larkin, on the board of the New York Zoological Society, pointed out that the American bison on their ranges are "pretty much managed rather than truly wild." Southeast Asian pheasants, reintroduced in their native habitat, he added, were soon shot out in exactly the same way that their predecessors had been in the first place.

Large carnivores present the most formidable problems of all.

Those who were inspired by *Born Free*, the story of how George and Joy Adamson returned Elsa to the wild, will also recall how extraordinarily difficult it was to get a hand-reared lioness to shift for herself in her new-found freedom. The Adamsons, by their own account, drove sixty miles to shoot a buck for Elsa because, having known only cut-up chunks of meat, she did not associate what she ate with living animals. The hungry lion repeatedly returned to the Adamsons, and once came crawling back suffering from a tick-borne virus against which as a household pet she had worked up no immunity, and which would no doubt have killed her if her human friends had not been on hand to nurse her back to health.

Warren Thomas points out an additional reason why, though it is possible to recondition a captive-bred animal, the job is so staggering. Such an individual has grown used to man's upright posture, normally intimidating to a wild animal, as well as to his odor and his sounds. Affected at any time by a shortage of prey, it naturally heads for the haunts of humans—with fatal consequences to them or to itself: "You've turned loose a time bomb."

Both he and Dr. Reed believe, on the other hand that in two or three generations of minimal human contact and, without even the slightest opportunity for imprinting, a captive animal might be psychologically ready for a shift to the wild. The question is whether there would be a territory for it to acquire. As Conway has pointed out, the 750 Siberian tigers now in zoos would require an area in nature about four times the size of Yellowstone National Park. That would be nearly 8.9 million acres, or 13,874 square miles—roughly the area of Massachusetts and Connecticut combined—for fewer than a thousand animals. Nor should it be forgotten that, in the words of K. S. Sankhala, director of Project Tiger, the same big cat "needs a wide range of prey, which, in its turn requires a wide spectrum of habitat. To ensure the survival of the tiger at the apex of this biomass, not only the animal itself, but the entire natural environment, must be preserved."

The case for zoos, then, does not rest on the possibility of reintroduction, but on all its other contributions—to science, to culture, to conservation itself—and on the simple fact that to have preserved *some* wild animals in good zoos is vastly better than having *no* animals, either in zoos or in the wild. On this point Henry Thoreau was surely right: "Every creature is better alive than dead, men and moose and pine trees, and he who understands it aright will rather preserve its life than destroy it."

Securing
the Safety Net

*For one species to mourn the death of another is a new thing under the
sun. The Cro-Magnon who slew the last mammoth thought only of steaks
. . . The sailor who clubbed the last auk thought of nothing at all. But
we, who have lost our last pigeons, mourn the loss.*

—ALDO LEOPOLD

Some of those animal admirers who reject zoos out of hand nevertheless
find highly desirable a widespread public appreciation of the majesty of
the elephant, the elegance of the gazelle, the awesomeness of the tiger.
They ask only that such appreciation be acquired on the spot, in the
far-off savannas of Africa, the jungles of India or Sumatra. Which is not
very different from advising a ghetto youth that a Harvard education
would be well worth his money.

Glorious as it is to go on a safari to the Serengeti, or ride an elephant
in the forest park of Kanha, the opportunity for such adventuring is
open only to a minuscule fraction of the world's population. For an
American in 1980, a week's journey through a few of the animal parks
of East Africa called for an outlay of $2,500 on the skimpiest possible
budget, including round trip fare from New York—or $5,000 at the
outset for a vacationing couple. How many couples are there who can
afford to spend that on a seven-day vacation? Even the citizens of
Nairobi as a rule see little of their country's animal treasure. Aside from
an occasional bus ride arranged for school children to nearby Nairobi
Park, most of the city's residents, who can afford neither vehicles nor

guides, see less of the East African fauna in a lifetime than a normally energetic tourist sees in a day's safari.

The fact that more people visit the Bronx Zoo in the month of July than visit all the national parks of Africa in a year is not an argument, however, against the African parks. Their value for scientific observation, for esthetic appreciation, and for the propagation of species is enormous—so much so that zoo people are among their most ardent supporters. More than forty national parks, reserves, and sanctuaries around the world have been launched, encouraged, or supported in a major way by the New York Zoological Society alone. No zoo official I have encountered opposes the idea of international support for these havens of what is, after all, an international heritage. They are fully aware that the world's least developed countries, the habitat of most of its wildlife, are the ones that can least afford the investment in land, protection, personnel, and technology that is required for a great animal preserve—notwithstanding the money brought in by wildlife-minded tourists. If the rich and powerful United States can barely keep its remnant populations of wolf and grizzly bear, how can Indonesia be expected to support its orangutans or Zaire its gorillas?

And yet we know that all the world's national parks together represent an insignificant part of the earth's land surface—only one per cent of all its 55 million square miles—and contain only a minority of its threatened species. We know, too, that unless there is an unprecedented upsurge in the world's awareness of the problem, support by the rich nations is not to be counted on. At best, we have here not a choice but a crying need to maintain species as long as we possibly can *both in the wild, including the animal preserves, and in captivity.* Lord Zuckerman's conclusion seems irrefutable: "If I were as certain that we could preserve Przewalski's horse in the wild as we can in captivity, I would prefer to see it in the wild—but as I am not certain, I would rather have it in both."

If American zoo directors are not on the stump pleading for federal funds to sustain the national parks of other countries, it is primarily because they are busy pleading with local governments, their own private backers, and the public in general for the money they need to do a decent job of running their own establishments. For all public and quasi-public institutions devoted to the arts and sciences, the beggar's cup is always out; but the zoos' need for help is in several ways the most unrelenting, and its sources are the most limited.

A commercial zoo operates of course for profit. To entice an adequate attendance, a private enterprise such as the Busch Gardens in Florida offers itself as a combination zoo, vaudeville show, carnival, and amusement park. However good the animal collection is, the whole project is there to make money for Anheuser–Busch Inc. and to promote its beer. When it ceases to serve that purpose, it will disappear, and its excellent collection of animals, not to mention its capable and devoted staff, will move on—to whatever degree there is room for them to move in the competitive world of zoos.

Nonprofit establishments, even some of the very good ones, may similarly depend on sideshows to support the serious work of studying and breeding endangered animals. In contrast, museums of art and of natural science, both of which regularly receive government largesse, would scorn the idea of selling peanuts with Picasso, much less of luring visitors to their doors with dancers got up to suggest a chorus line of Cro-Magnon girls.

Yet the economics of zoo-keeping are, if anything, more rigorously demanding than those of museums. Besides salaries, admittedly a paramount cost, the most pressing items in an inflationary period, to a degree unknown in comparable enterprises, are food, fuel, and construction. Special temperatures must be maintained throughout the year to give zoo inmates the diverse climates on which their lives depend—hot and humid for reptiles, refrigerated for the penguins. Fortunate and few, in this respect, are outdoor zoos, like San Diego's, blessed the year round with mild temperatures to which an extremely wide range of animals can readily adapt.*

The costs of zoo cuisine have of course kept pace with the increases in food prices for humans. In a great zoo like London's the animals eat annually more than 200 tons of hay, 90 tons of meat, 16 tons of fish, 130 tons of vegetables, and 19 tons of fruit, for a food bill of something between $200,000 and $250,000. At the Brookfield Zoo the walrus alone accounts for 45 to 60 pounds of mackerel and herring daily, scorning less expensive fish. The Bronx Zoo's food bill now comes to roughly $400,000 a year—and keeps going up, as it does everywhere else. Washington's National Zoo inhabitants ate $137,000 worth in fiscal 1972, $250,000 in 1978.

Even so, I was told by Dr. Boese, the director of the Milwaukee Zoo,

---

*According to William Donaldson, now director of the Philadelphia Zoo, the Cincinnati Zoo used to free its North American reptiles every fall rather than heat their building for the winter. It was cheaper to acquire new specimens in the spring.

"The public sees only between fifty and seventy-five cents of every dollar spent on a zoo." The rest is invested in barns, holding pens, transportation equipment, quarantine arrangements, and other outlays for the health and scientific observation of the animals. In the excitement over a glamorous new animal, the amount spent on its housing is easily overlooked. "The purchase of new attractions is a small part of the budget," David Hancocks points out. "Suppose you were lucky enough to get a couple of pandas free of charge, you couldn't take them unless you could afford a new facility that could cost as much as $850,000." It is the difficulty of getting funds for construction, with all that now goes into an acceptable replica of an animal's habitat, that so often prevents a mediocre zoo from rising to the status of a really good one.

It need hardly be said, of course, that a pair of giant pandas would not be free, and although the acquisition of new specimens is no longer at the top of zoo expenditures, it can still be a serious drain on the budget. As zoos find it less and less feasible to draw on the wild—even to the highly limited degree that is still permitted—rarity, regulations, and inflation combine to send up the prices of wild animals. If a zoo could buy an Indian one-horned rhinoceros at all (and that would have to be from another zoo), the price would be well over $50,000—as it would be for an okapi, or a pair of gorillas.

Sometimes it is not the rarity of the animal that raises the cost so much as the difficulty of rounding it up and getting it safely transported. That is why a gerenuk costs around $13,000. I was told at San Diego that a rare species of hartebeest could command as much as $70,000 a pair, and a bongo was quoted at a pair for $60,000. The little Thomson's gazelle, which can be seen by the thousands on the African plains, will cost a zoo about $1,800 and the even smaller muntjac about the same. The Bronx Zoo recently felt obliged to share with the National Zoo the ownership of three Elds deer, a rare Burmese species, rather than turn them down because of the $35,000 price tag. Among the birds, a penguin will run to well over $600 and a Chilean flamingo to $750.

The nearly extinct Asian lion recently cost the Marwell Zoo $12,000 a pair, but the African lion, which breeds so readily and eats so much, can be given away only if you are "lucky enough to find a home for him," says James Dolan, general curator of the San Diego Wild Animal Park. Even the tiger, whose kind are dying out in the wild, does so well in zoos that a male will hardly bring more than $500 in the zoo market, about one-seventh the price of a tiny lesser panda. No price can be quoted for the invaluable giant panda, nor for the pygmy chimpanzee, much less known than the panda but just as appealing.

Given the state of the animal market and the tightness of zoo bud-gets—particularly those dependent on municipal governments—loans of animals for breeding are bound to be increasingly favored. In the cooper-ative cultivation of a renewable resource, the lending and the receiving zoo have a common need to know all they can about the species in general and the transferred animal in particular. As a result a zoo director no longer has to buy a primate in a poke, as it were. Records and life histories of individual specimens are becoming the rule. Ameri-can zoos annually submit to the American Association of Zoological Parks and Aquariums (AAZPA) their inventories and, generally, a list of what animals are available for trading or lending, along with informa-tion as to age, pedigree, diet, and so on. Four zoos—the National, Bronx, Minnesota State, and San Diego—now employ full-time registrars to keep up with this flow of information.

The AAZPA sponsors a computer system (the International Species Inventory System, known as ISIS) which can show instantly which zoos in the country have, say, Siberian tigers, and the bloodlines of each, so as to avoid bringing in an animal too closely related to the one a zoo director hopes to breed. The *International Zoo Yearbook* publishes annual censuses and breeding records for rare animals in zoos. Various zoos around the world, moreover, keep stud books for animals in which they specialize. At Helsinki it is rare leopards, at Basel the Indian rhinoc-eros, at Whipsnade Père David deer, at Prague the Przewalski horse, at Antwerp the pygmy chimpanzee, and so forth. Skepticism has been voiced concerning the accuracy of some of these records, especially for the Przewalski, but the whole concept of scientific and worldwide record-keeping is very new and a far cry from the days when it was enough for zoos, each a land-based ark unto itself, to pair its animals off, two by two, and call it a breeding program.

Most of the world's great zoos are backed by societies made up of leading citizens whose chief function—by means of publicity, promotion, political influence, and public appeals—is to augment an income that must otherwise rest on the sale of tickets, souvenirs, food, and services, plus such aids as a municipality chooses to make in the way of land donations and tax concessions. Municipal and county zoos are subject to the ups and downs of city finances and the expediency of politicians. Los Angeles, a rich city, now does reasonably well by its zoo, though its record in the past was spotty at best. Municipalities hard pressed by rising social needs and declining revenues will hesitate to appropriate funds for a new sea lion's pool before they have met politically stringent demands for other public needs—which is to say, they will hesitate indefinitely.

The obvious recourse is to adopt the principle that all zoos that cannot maintain themselves adequately by their own efforts and from voluntary contributions should be closed as institutions that have not justified themselves by the iron law of supply and demand. But by that criterion we should soon have no museums and very little in the way of music, theater, dance, or any other form of human activity that costs more to produce with quality than individual patrons can afford to pay.

For contributions to the needs of humanity above the level of its daily bread, the National Endowment of the Arts, the National Endowment for the Humanities, and dozens of state councils properly spend millions of dollars in taxpayers' money every year, and the country is much the richer for it. Why not for zoos, which at their best present and preserve the works of nature as those other laudable institutions present the works of man?*

In view of the long-term investment zoos will have to make if wildlife is to be saved—an investment they cannot hope to recoup—their case for federal bounty is even stronger than that of many institutions that now receive it. Conway and Rabb recently worked out some illuminating figures concerning the 750 Siberian zoo tigers that already outnumber their relatives in the wild. Starting with the calculation that it costs $1,200,000 a year just to feed them, Dr. Conway reports:

> *Adding veterinary and curatorial care, heat, light, water, and building maintenance at a very conservative $4.50 per day per animal, the world zoo bill for annual Siberian tiger care is about $2,432,000. Disregarding capital improvements, sustaining a stable population of these felids until the year 2000 will cost at least $49 million. By contrast, rare Rembrandts in a museum will require only an occasional dusting.*

Besides dusting, art museums do have to worry about high insurance premiums, maintaining protective temperatures, lighting, occasional restorations, and other such items; but by and large there *is* something more durable and less demanding about a Gauguin than a gorilla.

*The National Endowment for the Arts and the National Endowment for the Humanities have contributed to zoos, but the grants have been extremely few and far between.

Should public funds, then, be spent on *all* zoos, regardless of quality and intent? Obviously not, anymore than a federal or state arts council should make grants without any distinction between a symphony orchestra with a record of achievement and a street musician cadging quarters. By no means are all of the five hundred or so animal collections in the United States entitled to so much as a dollar of the taxpayers' money. Hundreds of them are in no way entitled to call themselves zoos at all, certainly not those roadside affairs that use a few caged animals as a come-on for the sale of snacks and souvenirs. The same is true of those scores of tourist traps whose lurid billboards proclaim for five miles at hundred-yard intervals something like "Wild Congo Village—Fierce Jungle Beasts" and, like enough, show what appears to be King Kong chasing a blonde through the brush.

For bare sufferance, let alone government assistance, there must clearly be criteria. What makes a legitimate zoo? What qualifies it to hold individual creatures captive, no matter how justifiable some degree of captivity may be for the good of a species? From the zoo experts to whom I put these questions, or whose public comments on the subject I have read, the composite picture of a zoo deserving support would look something like this:

1. The animals are in good condition, physically and psychologically. "Look for an air of contentment," suggests Gerald Durrell, "glossy and tight feathering in birds; sleek, shiny coats on well-covered bodies in mammals; a healthy patina on reptiles, amphibians, and fish." When you find such signs, they are the consequences of proper zoo conditions— professional supervision, good diet, health care, enough space, and, more important, space planned and equipped to the individual needs of each species. You will see in such a collection no obviously neurotic behavior, such as constant pacing (not just before dinner is served), self-mutilation, and acts of obvious hostility.

2. The animals breed—always a sign that they are content with their surroundings and, so far as the purposes of the zoo go, a major reason for its existence. Obviously a single birth now and again does not qualify as breeding; only a consistent record of propagation, to the second and third generation, can be regarded as significant.

3. The zoo will be a cheerful, lively, and informative place—which is to say that it will be educating its patrons through well-planned exhibits, attractive graphics, and instructive programs, even while it stimulates their excitement in the natural world.

4. The inhabitants are studied scientifically, either by the zoo's per-

sonnel or by trained observers from the outside, not just to develop ways to improve their present lot but also to find out more about their ways than can be learned in the wild, where often enough their living habits are too hidden to be observed at all.

A criterion I have not seen or heard mentioned anywhere else is advanced by L. Dittrich, director of the zoo in Hannover, West Germany. Writing in *Animalia*, journal of the World Federation for the Protection of Animals, Dr. Dittrich describes "symbolic separations within a zoological garden that could be overcome without great effort by the animals kept behind them—if they wanted to," but which they show no desire to overcome. He specifies such jumping animals as antelopes, llamas, zebras, kangaroos, and some species of monkeys, all of which in Hannover are kept behind water-filled ditches only 1.9 to 2.5 meters wide. Since even the heavier antelopes can jump more than four meters and kangaroos as much as ten, the ditches are no barrier at all to animals that care to leave. "But once adjusted," he writes, these "zoo animals have no interest in leaving the area." The more such symbolic separations a zoo uses, the more likely that it is doing well by its charges.

Until the past decade, American zoos came under no national controls except for those governing the importing of animals, and such standards as states and municipalities imposed were few, variable, and vague. At the end of 1971 the Federal Animal Welfare Act went into effect, and although its enforcement by the Department of Agriculture leaves something to be desired—roadside zoos are still being indefensibly licensed—the law did lay down the proposition that the conditions of captive animals are a public concern. Two inspections a year are mandated to determine whether a collection meets certain minimal standards for feeding and care—cage size, sanitation, drinking water, and the like —but nothing as to broad purpose, the quality of the animals' lives, the professionalism of the staff, or the zoo's contributions either to the education of its patrons or the conservation of species.

It is only in the past few years that standards in these larger areas have been recognized in any concerted way by the zoo-keeping fraternity itself. The profession's arbiter is the American Association of Zoological Parks and Aquariums, to which all but a half dozen of the two hundred or so real zoos of the United States and Canada belong.

A zoo that seeks membership in the AAZPA must submit detailed information on all aspects of its establishment, educational plans, and philosophy, and be sponsored by three member zoos. The entire membership is invited to comment on an application, after which a thirteen-man board of directors votes on its admission. If granted, the applicant has two years to fulfill any requirements for accreditation that were not fully met at the time it was voted in.

More significantly, present member zoos will from now on have to submit to periodic review. Under a recently adopted rule, an inspecting team will call on each of them every three years to ascertain whether it is currently a credit to the association. If not, prescribed steps will be taken for further review and possible suspension. It should be possible in the near future, then, for the good zoos to be spared the handicap imposed on them by a label that has not always elicited admiration.

Ironically, it is the *least* profitable kind of animal collection that can assure the greatest improvement of zoos, offer most hope for the long-term rescue of threatened species, and give government a chance to make a highly significant and measurable contribution to the cause of conservation.

The burden of this book has been that the zoos' greatest service lies in the propagating of species that, for one reason or another, are fast fading from the earth. This is a service that calls for specialized collections, and requires those large wild animal parks—survival centers on the order of St. Catherine's Island and Front Royal—which without visitors are likewise without benefit of admission fees. Of least interest to a casual zoo-going public, moreover, they are least likely to attract the kind of philanthropy that rewards donors with the thanks and appreciation of the community.

To some extent, no doubt, such vital breeding preserves can be supported by consortiums of zoos. These could pool their contributions for a common benefit—the right to draw on a breeding center to restock their own collections and to enjoy through them the genetic variability necessary to keep their own smaller breeding operations vigorous. In the nature of things, however, the comparatively rich zoological societies, few in number, would have to carry the burden rather than the great number of hard-pressed budget-constricted zoos of the country, especially those dependent on equally hard-pressed city councils.

The hope, then, is that the federal government and the states—but principally the former—will see their way to the establishment and maintenance of regional wildlife centers, where threatened species could be propagated under the most biologically favorable circumstances. The idea has at least begun to sprout. In Ohio the zoos of Cleveland, Cincinnati, Columbus, and Toledo got together in the late 1970s to press for a state-financed breeding center. A bill to provide it was taken seriously in the legislature but failed for reasons having nothing to do with its merits, and since then members of Congress have shown a growing interest in the problem of endangered species.

Government assistance, should it come, would make possible the creation of statewide survival centers or, better still, regional ones appropriate to the country's varied climates. All the zoos of the world might in time draw on these breeding operations, the better to serve their function as a great safety net to break the fast-approaching fall of the wild.

Epilogue: The Intangibles

Not only men but all living things stand or fall together. Or rather man
is of all such creatures one of those least able to stand alone. If we think
only in terms of our own welfare, we are likely to find that we are losing
it.

—JOSEPH WOOD KRUTCH

If the practical uses of wildlife to humankind were the whole case for
trying to save threatened species, that case would still be sound, as I
hope to have demonstrated. But there is much more to the matter than
cost accounting, for in all those aspects of life that make man human,
the role of animals remains, as always, pervasive and profound.

To sense his continuing identity with the animal world, to feel again
something of our primitive ancestors' involvement with it, one must
almost recapture the vision of a child. Is it an accident that in even the
most urbanized cultures animal stories are the first, and the most endur-
ing, the eternally popular staple of childhood literature? Humans and
beasts not only communicate easily in those stories but casually move
in and out of each other's world like the fellow creatures they once were.

The young reader or listener is not misled by fantasies about sly
foxes, gentle fawns, and cheeky rabbits into thinking that the animal
world is truly filled with a whimsical quality—any more than he expects
a teddy bear to respond physically to his love. The pattern of wishful
anthropomorphism behind the whimsy is one that ought to drive no one
but a literalist to despair. For on a deeper level of that identification is

the truth itself: the child *is* an animal, and in his fantasy association with other animals he recapitulates a primeval relationship.

Psychologists and naturalists alike have made the point that the animal world provides small children with their first exercise in reducing their own sphere to manageable proportions by way of categories that begin to give it shape—the names of animals, the sounds they make ("What does the pig say? What does the cow say?"). Not long after come the human qualities that animals have been made to symbolize: the fearless lion, the bold eagle, the stubborn mule, and so on. It does not matter in the least that as natural history both sounds and attributes are far-fetched not to say preposterous. The point is that through them a child first learns to make distinctions and assign qualities. As the naturalist Paul Shepard writes in his striking and original book, *Thinking Animals:*

> *From physical traits, then behaviors, and on to whole chunks of familiar action, the animal is a kind of handle for abstractions. . . . As children get older, the stories they hear are increasingly interlaced with human figures. That is, human mothers and fathers appear in such tales quite early, then brothers and sisters, neighbors, friends, maidens and swains, soldiers and politicians. The repertoire of human passions and actions is slowly enlarged and repeated, clarified by its incarnation in specific animals, then transferred back to people and to the listening self.*

All this reflects a need, not of animals for man but rather of man for animals. Indeed what bedtime stories are to children, myths have been, from time beyond recording, to the whole of the human race— myths which do not merely include animals but frequently revolve about them, and about man's relations with them.

The pagan deities that figure in ancient mythologies spent much of their time either changing themselves into beasts the better to promote their designs, usually sexual, or changing mortals into animals, generally as an act of revenge. Zeus found it easier to carry off Europa when he assumed the guise of a white bull, and to seduce Leda by changing himself into a swan, just as his Roman counterpart changed Io into a heifer, hoping that way to conceal their dalliance from Juno, his divine but watchful spouse. None of this may show much concern with realism, but it does suggest a powerful sense of the unity of nature and of man's full and uninhibited partnership in it.

So did those marvelously sophisticated pictures on the walls of Altamira and Lascaux, which were the beginnings of a tradition of animal art that has continued unbroken for some 30,000 years. Likewise in the earliest works of the Egyptians, the Greeks, the Indians, the Chinese, the Cretans, and many other cultures, animals were omnipresent. And they continued to be so through the medieval bestiaries and the minutely detailed woodcuts of Dürer, right down to Landseer's haughty stags, the creatures of Rousseau's romantic wilderness, and the savagely twisted horses of Picasso.

The beasts of the field have been model, symbol, and inspiration—the embodiment of all that the poet or painter needed to set them to the work of depicting for their fellows the beauty, the terror, the sublimity of the natural order. The sight and sound of a skylark were all the "uses" of a bird that Shelley needed to know about, the tiger of Blake's poem all the evidence he required for his mystic view of creation. Even a layman with perhaps less exalted esthetic sense can be thrilled all the same by the magnificent curve of a Rocky Mountain sheep's horn, or the dazzling color of a quetzal. He can take pleasure in the sight of an impala sailing through the air like a stunt man fired from a cannon, the coiled tension in the crouch of a leopard (viewed of course at a safe distance), or the stately movement of giraffes across a plain. He can delight in the homely tenderness of a mother orangutan grooming her infant, or the self-important charm of king penguins, to all appearances caucusing like delegates at a political convention.

If it were not for this all but universal appeal—call it esthetic, psychological, or what you will—why have humans from the days of the Egyptians to this very moment flocked to zoos, menageries, animal shows of all kinds in such numbers? Why would a ship's passengers rush to the rail for even the distant sight of a whale, or a school of porpoises for that matter? Yet whale-watching has become so popular on both coasts that a single boat working out of Provincetown has booked 18,000 passengers in five years of catering to that interest. Why is a camera safari deemed worth all the effort and money it has cost at the moment when a real live leopard is sighted, four legs lazily drooped over the branch of a thorn tree? Why, for that matter, the flood of wildlife documentaries on television—good, bad, banal, or misleading—not to mention the countless large animal books that pile up on the coffee tables of the land?

Such are the complexities of human nature, however, that worship and pleasure are not always the predominant factors in the attitude of man to animal. There are also hatred, the will to subdue, even the deliberate inflicting of pain. Deploring these expressions of the human psyche, one may yet concede that they have played, and still play, a major part in the relationship of humans with their earthly companions. They can turn a man into the primitive combination of revenge and *machismo* that allows him to indulge in bullfighting and then in triumph to hold aloft his defeated opponent's ears and tail. And for one who has neither the opportunity nor the nerve to confront a wild beast on his own, the experience may be the vicarious one provided by books about a man-eating tiger in the India of the raj or the lion that fed regularly on African railway workers, allowing the reader the prospect that in the end, the great white hunter will at last do the monster in, as a grateful native population goes into a frenzied dance of rejoicing around the carcass.

In the Middle Ages trials of animals were common, complete with counsel, legal formalities, and public executions. By papal decree cats in the fifteenth century were burned at the stake along with the witches who kept them. As late as the eighteenth century, a cow was publicly hanged in France for some capital offense and soon afterward a female donkey escaped the same fate only through being pardoned on the basis of her previous good character.

At the opposite pole from such savagery toward animals is an equally anthropomorphic sentimentality—which, however innocent, has its own potential for mischief. The tendency of the sentimentalist is to sympathize with other animals in inverse proportion to a creature's physical differences from himself. A bulletin of the IUCN complains, very much to the point, that "many people who get dewy-eyed over seals would be very happy to see all crocodiles turned into handbags." The seals in turn get nothing comparable in the way of dewy-eyed attention to that bestowed on a bear cub or a baby gorilla. Indeed there is no zoo attraction as surefire as the suckling young of a mammal—the larger the mammal and the smaller the suckling, the better.

Lorus J. Milne and Margery Milne, biologists of the University of New Hampshire, have neatly tabulated this order of human sympathy:

We rationalize that human awareness exceeds that of any other animal, and that human suffering must be most intense, most worthy of continued help. Other mammals and birds follow, then cold-blooded vertebrates, next the invertebrates, of progressively smaller size and simpler organization, and, finally, the plants, both green and non-green. This hierarchy continues our traditional anthropocentricity, whereas reverence for life implies no rank order.

So much for those who can see the saving of whales but not of snail darters; for those who do not blink an eye at the feeding of a live cricket to a small mammal in the zoo, but shudder at the thought of letting even so distant a mammalian relative as a mouse be fed to a lowly reptile. So much, indeed, for those who will relish a lamb chop, wear cowhide on their feet, and pop a lobster into boiling water but will not hear of "imprisoning" an antelope in the paddock of a zoo or tolerate the shooting of a white-tailed deer in a region overpopulated by its kind.

This emotionalism over small, gentle, big-eyed creatures is referred to by the irreverent as the Bambi syndrome. The fact deserves mention as one more example in a chain of animal metaphors and similes that have marked man's language from the Bible to this morning's newspaper. No one suggests that wildlife should be saved because it enriches the language, but it is not far-fetched to say that if animals were truly remote from man, his speech would not resound with their presence.

It is ridiculously easy to draw up lists of everyday phrases that draw on animal life and lore—almost impossible, in fact, to imagine a healthy body of invective without them. Think of the stupidity condensed, rightly or not, in the epithet "jackass," of how satisfactorily the thought of licentiousness is conveyed by "old goat," or of cowardice by "jackal" or "chicken."

Can *Homo sapiens* really be as divorced as he sometimes pretends to be from the rest of the animal kingdom when his art, his literature, his history, his science, and his very language are filled to overflowing with allusions to fellow animals, from asses to zebras? Indeed, it is not as surely his own extinction that he has to fear in consequence of losing the earth's wildlife, as it is the drabness of a world without those creatures to which he is so tied by emotion and by their long joint history.

Over and over again, in the course of preparing this book, I asked people devoted to the saving of animals why it all mattered. Few of them cited the practical services of animal to man, great as those are. Overwhelmingly they talked of the color, the beauty, the excitement that would go out of the world—of the loss to civilization, should the day come when nowhere on this planet can a human being look upon a living lion, an elephant, a leopard, or a whale; when a rhinoceros, like the brontosaurus, is no more than a picture in a book; when the giraffe survives only as another plaster cast to be hurried by in a museum.

Some were more eloquent than others, but no one, I think, surpassed the often-quoted William Beebe, who in his days as curator of birds at the Bronx Zoo made the cultural case for saving endangered species: "The beauty and genius of a work of art may be reconceived, though its first material expression be destroyed; a vanished harmony may yet again inspire the composer; but when the last individual of a race of living things breathes no more, another heaven and another earth must pass before such a one can be again."

Acknowledgments

In seeking information for this book, sounding the experience of people for whom a study of animals is the dedication of a lifetime, I visited some forty zoos and wildlife reserves in North and Central America, Europe, Africa, and India. Usually I was able to make a satisfying study, sometimes over a period of several days and for much of the time in the company of an obliging director, curator, head keeper, educational officer, veterinary, or a whole team of such experts. Everywhere I went I was treated with a friendliness that belies the old canard that the more people care for animals, the less concern they have for other people. My guides and those whom I interviewed were unfailingly cordial, considerate, and helpful.

The zoos and animal parks on my itinerary, and the staff members who principally and patiently responded to my curiosity were the following: *New York Zoological Society (Bronx Zoo)*, William G. Conway, general director; James G. Doherty, Joseph Bell, Richard L. Lattis. *Wildlife Survival Center*, St. Catherine's Island, Georgia, John A. Lukas, associate curator; Frank Larkin and June Noble Larkin, John Woods. *National Zoological Park (Smithsonian Institution)*, Washington,

D.C., Theodore Reed, director; John Eisenberg, Devra Kleiman, Jaren Horsley, Judith White. *Conservation and Research Center,* Front Royal, Virginia, Larry Collins, curator of mammals.

Arizona–Sonora Desert Museum, near Tucson, Arizona, Dan Davis, director; Inge Poglayen. *San Diego Zoo,* Clayton Swanson, general manager. *San Diego Wild Animal Park,* James Dolan, general curator. *Los Angeles Zoo,* Warren D. Thomas, director; Ed Alonso. *Woodland Park Zoo,* Seattle, Washington, David Hancocks, director. *Lincoln Park Zoo,* Chicago, Illinois, Lester E. Fisher, director; Kevin S. Bell, Edward Almandarz. *Brookfield Zoo,* near Chicago, George Rabb, director. *Milwaukee County Zoo,* Milwaukee, Wisconsin, Gilbert Boese, director.

Also, *Philadelphia Zoo,* Philadelphia, Pennsylvania, William Donaldson, director, Harold Bair; Robert L. Snyder, director, Penrose Research Laboratory. *Catskill Game Ranch,* Ocala, Florida, Roland and Kathryn Lindemann, owners; Claus Reitzer. *Busch Gardens,* Tampa, Florida, Gerald Lentz, manager, zoo department. *Zürich Zoological Garden,* Zürich, Switzerland, Peter Weilenmann, director; R. E. Honegger. *Basel Zoo,* Basel, Switzerland, Hans Wackernagel, vice-director. *Frankfurt Zoo,* Frankfurt, West Germany, Richard Faust, director. *Blijdorp Zoo,* Rotterdam, Netherlands, D. Van Dam, director. *Wassenaar Zoo,* Wassenaar, Netherlands, Jan W. W. Louwman, director.

Also, *Zoological Society of London,* Lord Zuckerman, secretary; Colin Rawlins, director of zoos, including *London Zoo. Whipsnade Park,* near Dunstable, England, V. J. A. Manton, curator; Owen Chamberlain. *Marwell Zoological Park,* Winchester, England, J. M. Knowles, director.

A number of other zoos that I had visited in recent years, although before I started on this work, included, besides quite a few smaller collections, those in Pittsburgh, Baltimore, Toronto, and Chester (England). I greatly regret having had to miss others I wanted to see—the municipal zoos of Mexico City, St. Louis, and the Gladys Porter Zoo in Brownsville, Texas—but the aim had necessarily to be representative rather than all-inclusive.

To these tours of close to thirty zoos, let me add the trips I made —as a member of the New York *Times* Editorial Board concerned chiefly with environmental matters—to the great animal preserves of Africa and India. These journeys were not generally made on direct assignment, though they yielded articles and editorials for the *Times;* but it was my position on that newspaper that made them possible, and I want to record here my debt to John B. Oakes, then chief of the editorial page, for encouraging me to pursue an interest in wild animals that had been with me from early childhood.

On these trips I was able to see something of the remaining wildlife of Kenya, at Lake Nakuru, Nairobi Park, Masai Mara, and the Aberdares; of Tanzania, at Ngorongoro Crater and on the Serengeti; and of India at the Bharatpur bird sanctuary, Corbett and Kanha parks, the Sariska Wildlife Sanctuary, and, best of all, at Kaziranga, near the Chinese and Nepalese borders of Assam. Subsequently I also had the opportunity of traveling to the North Slope of Alaska and of wandering through rain and cloud forests in Central America and the Caribbean.

These journeys were made all the more rewarding by the guidance of those I interviewed and the wisdom I gleaned from the speeches, reports, and books of those whom I regrettably did not. Among the authorities not connected with zoos with whom I profitably talked were Norman Myers, the distinguished conservation scientist, at Nairobi; Robinson McIlvaine, of the African Wildlife Leadership Association, who also loaned me a jeep and driver to Lake Nakuru; John Hopcraft, who guided me through that marvelous flamingo capital; Gerardo Budowski, head of the Renewable Resources Program of the Tropical Agricultural Research and Training Center at Turrialba, Costa Rica; K. S. Sankhala, the moving spirit of Project Tiger; Thomas Lovejoy, the World Wildlife Fund's United States vice-president for science; and Lee Talbot, formerly conservation director of the fund and now director general of the International Union for the Conservation of Nature.

Also Paul Opler, of the Department of the Interior's Office of Endangered Species; Jon Tinker and Kath Adams, of Earthscan in London; Karl Frucht, administrator of the World Federation for the Protection of Animals; John W. Grandy, head of Defenders of Wildlife; Sue Pressman, of the Humane Society of the United States; Earl Baysinger, then with the IUCN; David Mack, of the U.S. office of TRAFFIC (Trade Records Analysis of Flora and Fauna in Commerce); Elvis Stahr, Washington representative of the National Audubon Society, who made possible my travels to India; and Barbara A. Hockfield, formerly with the education department of the Roger Williams Zoo in Providence, Rhode Island.

I owe a very special debt to Russell Train and the World Wildlife Fund, and to the Natural Resources Defense Council, by way of Larry Rockefeller, for grants that helped make it possible for me to do the travel essential to preparing this book. Above all, I must express profound gratitude to William G. Conway. Beyond opening to me the workings of his own Bronx Zoo, Mr. Conway generously paved my way with letters of introduction to zoo personnel here and abroad, and still more generously read the manuscript of this book, setting me straight on scientific inaccuracies. For whatever such errors remain, the responsibility can only be mine, either because I inadvertently failed to correct them or in doing so compounded the felonies.

Finally, what am I to say but many thanks to my wife, Kas, who not only zeroed in on hazy passages, asking helpful questions, but who typed and typed—and is still typing as I write this in heartfelt appreciation.

Selected Bibliography

BOOKS

AMORY, CLEVELAND. *Man Kind?* New York: Harper & Row, 1974.

BRIDGES, WILLIAM. *Gathering of Animals.* New York: Harper & Row, 1974.

BURTON, ROBERT. *The Mating Game.* New York: Crown Publishers, Inc., 1976.

CAMPBELL, SHELDON. *Lifeboats to Ararat.* New York: New York Times Books, 1978.

CARAS, ROGER. *Dangerous to Man.* New York: Holt, Rinehart & Winston, 1977.

CARRIGHAR, SALLY. *Wild Heritage.* Boston: Houghton Mifflin Co., 1965.

CLARK, KENNETH. *Animals and Men.* New York: William Morrow & Co., 1977.

CURRY-LINDAHL, KAI. *Let Them Live.* New York: William Morrow & Co., 1972.

DOUGLAS-HAMILTON, IAIN and ORIA. *Among the Elephants.* New York: Viking Press, 1975.

DURRELL, GERALD. *A Zoo in My Luggage.* Middlesex, England: Penguin, 1964.

DURRELL, GERALD. *The Stationary Ark.* New York: Simon & Schuster, 1976.

ECKHOLM, ERIK. *Disappearing Species: the Social Challenge,* Worldwatch Paper 22. Washington: Worldwatch Institute, 1978.

FISHER, JAMES. *Zoos of the World, the Story of Animals in Captivity.* Garden City, N.Y.: Natural History Press, 1967.

GOODALL, JANE. *In the Shadow of Man.* Boston: Houghton Mifflin Co., 1971.

GRZIMEK, BERNHARD. *Serengeti Shall Not Die.* New York: E. P. Dutton, 1962.

GRZIMEK, BERNHARD, editor in chief. *Grzimek's Animal Life Encyclopedia,* 13 volumes. New York: Van Nostrand Reinhold Co., 1972–1975.

HAHN, EMILY. *Animal Gardens.* Garden City, N. Y.: Doubleday & Co., 1967.
HAYES, HAROLD P. *The Last Place on Earth.* New York: Stein & Day, 1977.
HEDIGER, H. *The Psychology and Behavior of Animals in Zoos and Circuses.* New York: Dover Publications, Inc., 1978.
LAYCOCK, GEORGE. *The Alien Animals, the Story of Imported Wildlife.* New York: Ballantine Books, 1970.
LEAKEY, RICHARD E. and LEWIN, ROGER. *Origins.* New York: E. P. Dutton, 1977.
LEY, WILLY. *The Lungfish, the Dodo, and the Unicorn.* New York: Viking Press, 1948.
LEY, WILLY. *Dragons in Amber.* New York: Viking Press, 1951.
LIVINGSTON, BERNARD. *Zoo—Animals, People, Places.* New York: Arbor House, 1974.
MATTHIESSEN, PETER. *The Tree Where Man Was Born.* New York: E. P. Dutton & Co., 1972.
MEYER, ALFRED, editor. *A Zoo for All Seasons.* Washington: Smithsonian Institution, 1979.
MOWAT, FARLEY. *Never Cry Wolf.* New York: Dell, 1965.
MYERS, NORMAN. *Long African Day.* New York: Macmillan, 1972.
MYERS, NORMAN. *The Sinking Ark.* New York: Pergamon Press, 1979.
PFEIFFER, JOHN E. *The Emergence of Man.* New York: Harper & Row, 1969.
REGENSTEIN, LOUIS. *The Politics of Extinction.* New York: Macmillan, 1975.
RICCIUTI, EDWARD R. *Killer Animals.* New York: Walker & Co., 1976.
SCHALLER, GEORGE. *Year of the Gorilla.* New York: Ballantine Books, 1964.
SHEPPARD, PAUL. *Thinking Animals.* New York: Viking Press, 1978.

PERIODICALS

I have found extremely useful the annual reports of the Zoological Society of London (and likewise its scientific reports), of the New York Zoological Society, and of the National Zoological Park. Mines of information are the files of *International Wildlife, National Wildlife, Audubon Magazine,* the *Conservation Foundation Letter, Conservation Report, NRDC Letter* (National Resources Defense Council), *Sierra Club Bulletin, Defender's Magazine,* both *Animalia* and *Animal Regulation Studies* (World Federation for the Protection of Animals), *World Wildlife News,* and *Focus* (World Wildlife Fund—U.S.) as well as publications of Earthscan (International Institute for Environment and Development). Not least, I recommend the guidebooks and periodicals—particularly *Animal Kingdom* and *ZooNooz*—of many of the zoos discussed in this volume.

Appendix

Following is a list of the world's most threatened animals as listed in the *Red Data Book* of the International Union for the Conservation of Nature and Natural Resources. The categories, determined by the Union's Survival Service Commission, are endangered, vulnerable, and rare. *Endangered* animals are those "in danger of extinction and whose survival is unlikely if the causal factors continue operating." *Vulnerable* are those "believed likely to come into the endangered category in the near future if the causal factors continue operating." *Rare* are those "with small world populations that are not at present endangered or vulnerable, but are at risk."

	ENDANGERED	VULNERABLE	RARE
MAMMALS			
Primates	Black lemur (*Lemur macaco macaco*)	Mongoose lemur (*Lemur mongoz*)	Nossi-bé sportive lemur (*Lepilemur mustelinus dorsalis*)
	Red-fronted lemur (*Lemur macaco rufus*)	Grey gentle lemur (*Hapalemur griseus*)	Broad-nosed gentle lemur (*Hapalemur simus*)
	Sclater's lemur (*Lemur macaco flavifrons*)	Fat-tailed dwarf lemur (*Cheirogaleus medius*)	Hairy-eared dwarf lemur (*Allocebus trichotis*)
	Sanford's lemur (*Lemur macaco sanfordi*)	Coquerel's mouse lemur (*Microcebus coquereli*)	Perrier's sifaka (*Propithecus diadema perrieri*)
	Red-tailed sportive lemur (*Lepilemur mustelinus ruficaudatus*)	Western woolly avahi (*Avahi laniger occidentalis*)	Uhehe red colobus (*Colobus badius gordonorum*)

White-footed sportive lemur (Lepilemur mustelinus leucopus)

Indris (Indri indri)

Verreaux's sifaka (Propithecus verreauxi)

Aye-aye (Daubentonia madagascariensis)

Philippine tarsier (Tarsius syrichta)

Buff-headed marmoset (Callithrix flaviceps)

Cotton-top tamarin (Saguinus oedipus oedipus)

Golden tamarin (Leontopithecus rosalia)

Central American squirrel (Saimiri oerstedi)

Woolly spider monkey (Brachyteles arachnoides)

Yellow-tailed woolly monkey (Lagothrix flavicauda)

Tana River mangabey (Cercocebus galeritis galeritis)

White-eared marmoset (Callithrix aurita)

White marmoset (Callithrix argentata leucippe)

White-nosed saki (Chiropotes albinasus)

Black-headed uakari (Cacajao melanocephalus)

Woolly monkey (Lagothrix lagothricha)

Barbary macaque (Macaca sylvana)

Black colobus (Colobus satanas)

Proboscis monkey (Nasalis larvatus)

Kloss's gibbon (Hylobates klossi)

Chimpanzee (Pan troglodytes)

Pygmy chimpanzee (Pan paniscus)

Gorilla (Gorilla gorilla)

Zanzibar red colobus (Colobus badius kirkii)

Golden langur (Presbytis geei)

Olive colobus (Colobus verus)

Snub-nosed langur (Rhinopithecus roxellanae)

ENDANGERED	VULNERABLE	RARE
	Nilgiri langur (*Presbytis johnii*)	
Drill (*Papio leucophaeus*)		
Lion-tailed macaque (*Macaca silenus*)		
Tana River red colobus (*Colobus badius rufomitratus*)		
Preuss's red colobus (*Colobus badius preussi*)		
Pig-tailed langur (*Simias concolor*)		
Douc langur (*Pygathrix nemaeus*)		
Pileated gibbon (*Hylobates pileatus*)		
Javan gibbon (*Hylobates moloch*)		
Orang-utan (*Pongo pygmaeus*)		
Mountain gorilla (*Gorilla gorilla beringei*)		

Edentata

	Giant anteater (*Myrmecophaga tridactyla*)	Maned sloth (*Bradypus torquatus*)

Lagomorpha	Ryukyu rabbit (*Pentalagus furnessi*)	Lesser pichiciego (*Chlamyphorus truncatus*)	Greater pichiciego (*Burmeisteria retusa*)
	Volcano rabbit (*Romerolagus diazi*)	Giant armadillo (*Priodontes giganteus*)	Sumatra short-eared rabbit (*Nesolagus netscheri*)
	Assam rabbit (*Caprolagus hispidus*)		
Rodentia	Delmarva Peninsula fox squirrel (*Sciurus niger cinereus*)	Menzbier's marmot (*Marmota menzbieri*)	Utah prairie dog (*Cynomys parvidens*)
	Morro Bay kangaroo rat (*Dipodomys heermanni morroensis*)	Mexican prairie dog (*Cynomys mexicanus*)	Big-eared kangaroo rat (*Dipodomys elephantinus*)
	Salt-marsh harvest mouse (*Reithrodontomys raviventris*)	Chinchilla (*Chinchilla laniger*)	Texas kangaroo rat (*Dipodomys elator*)
		Jamaican hutia (*Geocapromys brownii*)	Block Island meadow vole (*Microtus pennsylvanicus provectus*)
			Beach meadow vole (*Microtus breweri*)

	ENDANGERED	VULNERABLE	RARE
			Thin-spined porcupine (*Chaetomys subspinosus*)
			Bushy-tailed hutia (*Capromys melanurus*)
			Dwarf hutia (*Capromys nanus*)
			Bahaman hutia (*Geocapromys ingrahami*)
			Cuvier's hutia (*Plagiodontia aedium*)
			Dominican hutia (*Plagiodontia hylaeum*)
Cetacea	Indus dolphin (*Platanista indi*)	Northern bottlenose whale (*Hyperoodon ampullatus*)	
	Blue whale (*Balaenoptera musculus*)	Vaquita (*Phocoena sinus*)	
	Humpback whale (*Megaptera novae-angliae*)	Fin whale (*Balaenoptera physalus*)	
	Bowhead whale (*Balaena mysticetus*)		
	Black right whale (*Eubalaena glacialis*)		

Carnivora

Northern Rocky Mountain wolf (*Canis lupus irremotus*)	Wolf (*Canis lupus*)	Small-eared dog (*Atelocynus microtis*)
Red wolf (*Canis rufus*)	Maned wolf (*Chrysocyon brachyurus*)	Bush dog (*Speothos venaticus*)
Simien fox (*Canis simensis*)	Asiatic wild dog (*Cuon alpinus*)	Barren-ground grizzly bear (*Ursus arctos richardsoni*)
Northern kit fox (*Vulpes velox hebes*)	African wild dog (*Lycaon pictus*)	Giant panda (*Ailuropoda melanoleuca*)
Baluchistan bear (*Selenarctos thibetanus gedrosianus*)	Spectacled bear (*Tremarctos ornatus*)	Celebes giant palm civet (*Macrogalidia musschenbroeki*)
Mexican grizzly bear (*Ursus arctos nelsoni*)	Polar bear (*Ursus maritimus*)	Turkmenian caracal lynx (*Felis caracal michaelis*)
Black-footed ferret (*Mustela nigripes*)	European otter (*Lutra lutra*)	Bornean bay cat (*Felis badia*)
Marine otter (*Lutra felina*)	Malagasy civet (*Fossa fossa*)	Andean cat (*Felis jacobita*)
La Plata otter (*Lutra platensis*)	Fanalouc (*Eupleres goudotii*)	
Southern river otter (*Lutra provocax*)	Fossa (*Cryptoprocta ferox*)	
Giant otter (*Pteronura brasiliensis*)	Brown hyaena (*Hyaena brunnea*)	
Cameroon clawless otter (*Aonyx microdon*)	Ocelot (*Felis pardalis*)	

ENDANGERED	VULNERABLE	RARE
Malabar large-spotted civet (*Viverra megaspila civettina*)	Clouded leopard (*Neofelis nebulosa*)	
Barbary hyaena (*Hyaena hyaena barbara*)	Leopard (*Panthera pardus*)	
Spanish lynx (*Felis pardina*)	Jaguar (*Panthera onca*)	
Pakistan sand cat (*Felis margarita scheffeli*)	Cheetah (*Acinonyx jubatus*)	
Eastern cougar (*Felis concolor cougar*)		
Florida cougar (*Felis concolor coryi*)		
Iriomote cat (*Prionailurus iriomotensis*)		
Asiatic lion (*Panthera leo persica*)		
Tiger (*Panthera tigris*)		
Barbary leopard (*Panthera pardus panthera*)		
South Arabian leopard (*Panthera pardus nimr*)		
Anatolian leopard (*Panthera pardus tulliana*)		

	Amur leopard (*Panthera pardus orientalis*)		
	Sinai leopard (*Panthera pardus jarvisi*)		
	Snow leopard (*Panthera uncia*)		
	Asiatic cheetah (*Acinonyx jubatus venaticus*)		
Pinnipedia	Japanese sea lion (*Zalophus californianus japonicus*)	Galápagos fur seal (*Arctocephalus galapagoensis*)	Saimaa seal (*Phoco hispida saimensis*)
	Mediterranean monk seal (*Monachus monachus*)	Juan Fernandez fur seal (*Arctocephalus philippii*)	
	Caribbean monk seal (*Monachus tropicalis*)	Guadalupe fur seal (*Arctocephalus townsendi*)	
	Hawaiian monk seal (*Monachus schauinslandi*)		
Proboscidae		African elephant (*Loxodonta africana*)	
		Asian elephant (*Elephas maximus*)	
Sirenia	Amazonian manatee (*Trichechus inunguis*)	Dugong (*Dugong dugong*)	

ENDANGERED	VULNERABLE	RARE
	Caribbean manatee (*Trichechus manatus*)	
	West African manatee (*Trichechus senegalensis*)	
Perissodactyla		
Przewalski's horse (*Equus przewalskii*)	Asiatic wild ass (*Equus hemionus*)	
Indian wild ass (*Equus hemionus khur*)	Mountain zebra (*Equus zebra*)	
Syrian wild ass (*Equus hemionus hemippus*)	Grévy's zebra (*Equus grevyi*)	
African wild ass (*Equus asinus*)	Black rhinoceros (*Diceros bicornis*)	
Mountain tapir (*Tapirus pinchaque*)		
Central American tapir (*Tapirus bairdii*)		
Malayan tapir (*Tapirus indicus*)		
Great Indian rhinoceros (*Rhinoceros unicornis*)		
Javan rhinoceros (*Rhinoceros sondaicus*)		
Sumatran rhinoceros (*Didermoceros sumatrensis*)		

Northern square-lipped
rhinoceros (*Ceratotherium
simum cottoni*)

Artiodactyla			
	Pygmy hog (*Sus salvanius*)	Babirusa (*Babyrousa babyrussa*)	Kuhl's deer (*Axis kuhli*)
	Fea's muntjac (*Muntiacus feae*)	Chacoan peccary (*Catagonus wagneri*)	Golden takin (*Budorcas taxicolor bedfordi*)
	Persian fallow deer (*Dama mesopotamica*)	Pygmy hippopotamus (*Choeropsis liberiensis*)	Hunter's hartebeest (*Damaliscus hunteri*)
	Swamp deer (*Cervus duvauceli*)	Vicuña (*Lama vicugna*)	
	Manipur brow-antlered deer (*Cervus eldi eldi*)	Wild Bactrian camel (*Camelus bactrianus*)	
	Thailand brow-antlered deer (*Cervus eldi siamensis*)	Himalayan musk deer (*Moschus moschiferus moschiferus*)	
	Formosan sika (*Cervus nippon taiouanus*)	Calamian deer (*Axis calamianensis*)	
	Ryukyu sika (*Cervus nippon keramae*)	North Andean huemul (*Hippocamelus antisensis*)	
	North China sika (*Cervus nippon mandarinensis*)	Marsh deer (*Blastocerus dichotomus*)	
	Shansi sika (*Cervus nippon grassianus*)	Asiatic buffalo (*Bubalus bubalis*)	

ENDANGERED	VULNERABLE	RARE
South China sika (*Cervus nippon kopschi*)	Gaur (*Bos gaurus*)	
Corsican red deer (*Cervus elaphus corsicanus*)	Banteng (*Bos javanicus*)	
Shou (*Cervus elaphus wallichi*)	Lechwe (*Kobus leche*)	
Barbary deer (*Cervus elaphus barbarus*)	Scimitar-horned oryx (*Oryx dammah*)	
Hangul (*Cervus elaphus hanglu*)	Addax (*Addax nasomaculatus*)	
Yarkand deer (*Cervus elaphus yarkandensis*)	Beira antelope (*Dorcatragus megalotis*)	
Bactrian or Bokharan deer (*Cervus elaphus bactrianus*)	Dibatag (*Ammodorcas clarkei*)	
Columbia white-tailed deer (*Odocoileus virginianus leucurus*)	Nilgiri tahr (*Hemitragus hylocrius*)	
Cedros Island deer (*Odocoileus hemionus cerrosensis*)	Markhor (*Capra falconeri*)	
Argentine Pampas deer (*Ozotoceros bezoarticus celer*)	Bighorn sheep (*Ovis cañadensis*)	

Lower California pronghorn
(*Antilocapra americana peninsularis*)

Sonoran pronghorn
(*Antilocapra americana sonoriensis*)

Western giant eland
(*Taurotragus derbianus derbianus*)

Tamaraw (*Bubalus mindorensis*)

Lowland anoa (*Bubalus depressicornis*)

Mountain anoa (*Bubalus quarlesi*)

Kouprey (*Bos sauveli*)

Wild yak (*Bos grunniens*)

Jentink's duiker (*Cephalophus jentinki*)

Giant sable antelope
(*Hippotragus niger variani*)

Arabian oryx (*Oryx leucoryx*)

Tora hartebeest (*Alcelaphus buselaphus tora*)

ENDANGERED	VULNERABLE	RARE
Swayne's hartebeest (*Alcelaphus buselaphus swaynei*)		
Zanzibar suni (*Nesotragus moschatus moschatus*)		
Black-faced impala (*Aepyceros melampus petersi*)		
Sand gazelle (*Gazella subgutturosa marica*)		
Moroccan dorcas gazelle (*Gazella dorcas massaesyla*)		
Saudi Arabian gazelle (*Gazella dorcas saudiya*)		
Pelzeln's gazelle (*Gazella dorcas pelzelni*)		
Arabian gazelle (*Gazella gazella arabica*)		
Cuvier's gazelle (*Gazella cuvieri*)		
Slender-horned gazelle (*Gazella leptoceros*)		
Mhorr gazelle (*Gazella dama mhorr*)		

Rio de Oro dama gazelle (*Gazella dama lozanoi*)

Sumatran serow (*Capricornis sumatraensis sumatraensis*)

Arabian tahr (*Hemitragus jayakari*)

Walia ibex (*Capra walie*)

Pyrenean ibex (*Capra pyrenaica pyrenaica*)

Straight-horned markhor (*Capra falconeri megaceros*)

Mediterranean mouflon (*Ovis ammon musimon*)

Marsupialia

Bridle nail-tailed wallaby (*Onychogalea fraenata*)

Crescent nail-tailed wallaby (*Onychogalea lunata*)

Leadbeater's possum (*Gymnobelideus leadbeateri*)

Pig-footed bandicoot (*Chaeropus ecaudatus*)

Rabbit-eared bandicoot (*Macrotis lagotis*)

Eastern jerboa marsupial (*Antechinomys laniger*)

Yellow-footed rock wallaby (*Petrogale xanthopus*)

Parma wallaby (*Macropus parma*)

Western hare-wallaby (*Lagorchestes hirsutus*)

Banded hare-wallaby (*Lagostrophus fasciatus*)

Desert rat-kangaroo (*Caloprymnus campestris*)

Northern rat-kangaroo (*Bettongia tropica*)

Lesueur's rat-kangaroo (*Bettongia lesueur*)

	ENDANGERED	VULNERABLE	RARE
	Thylacine (Thylacinus cynocephalus)		Scaly-tailed possum (Wyulda squamicaudata)
			Red-tailed phascogale (Phascogale calura)
			Dibbler (Antechinus apicalis)
			Narrow-nosed planigale (Planigale tenuirostris)
			Kimberley planigale (Planigale subtilissima)
Insectivora	Haitian solenodon (Solenodon paradoxus)	Mindanao gymnure (Podogymnura truei)	Cuban solenodon (Atopogale cubana)
		Russian desman (Desmana moschata)	Juliana's golden mole (Amblysomus julianae)
			Giant golden mole (Chrysopalax Trevelyani)
			Maquassi musk-shrew (Crocidura maquassiensis)
			Pyrenean desman (Galemys pyrenaicus)
Chiroptera	Rodrigues flying fox (Pteropus rodricensis)	Marianas flying fox (Pteropus mariannus)	Mauritian flying fox (Pteropus niger)
	Guam flying fox (Pteropus tokudae)	Indiana bat (Myotis sodalis)	Hawaiian hoary bat (Lasiurus cinereus semotus)

Ozark big-eared bat (Corynorhinus townsendii ingens)

Ghost bat (Macroderma gigas)

Singapore roundleaf horseshoe bat (Hipposideros ridleyi)

Gray bat (Myotis grisescens)

AMPHIBIANS

Caudata

Japanese giant salamander (Andrias japonicus)

California tiger salamander (Ambystoma tigrinum californiense)

Santa Cruz long-toed salamander (Ambystoma macrodactylum croceum)

Lake Patzcuaro salamander (Ambystoma dumerili dumerili)

Gold-striped salamander (Chioglossa lusitanica)

Desert slender salamander (Batrachoseps aridus)

Lake Lerma salamander (Ambystoma lermaensis)

Jemez Mountain salamander (Plethodon neomexicanus)

Texas blind salamander (Typhlomolge rathbuni)

Axolotl (Ambystoma mexicanum)

Olm (Proteus anguinus)

Kern Canyon slender salamander (Batrachoseps simatus)

Tehachapi slender salamander (Batrachoseps stebbinsi)

Shasta salamander (Hydromantes shastae)

	ENDANGERED	VULNERABLE	RARE
			Limestone salamander (*Hydromantes brunus*)
			Red Hills salamander (*Phaeognathus hubrichti*)
			San Marcos salamander (*Eurycea nana*)
Salientia	Israel painted frog (*Discoglossus nigriventer*)	Cape platana (*Xenopus gilli*)	Coromandel leiopelma (*Leiopelma archeyi*)
	Italian spade-foot toad (*Pelobates fuscus insubricus*)	Platypus frog (*Rheobatrachus silus*)	Stephen Island leiopelma (*Leiopelma hamiltoni*)
	Houston toad (*Bufo houstonensis*)	Sonoran green toad (*Bufo retiformis*)	North Island leiopelma (*Leiopelma hochstetteri*)
	Golden toad (*Bufo periglenes*)	Black or Inyo County toad (*Bufo exsul*)	Golden Coqui frog (*Eleutherodactylus jasperi*)
	Vegas Valley leopard frog (*Rana pipiens fisheri*)	Mt. Nimba viviparous toad (*Nectophrynoides occidentalis*)	Baw Baw frog (*Philoria frosti*)
		Goliath frog (*Conrana goliath*)	Pine Barrens tree frog (*Hyla andersoni*)

REPTILES

| *Testudines* | | Central American river turtle (*Dermatemys mawii*) | |

Crocodylia

Chinese alligator (*Alligator sinensis*)

Spectacled caiman (*Caiman crocodilus crocodilus*)

Rio Apaporis caiman (*Caiman crocodilus apaporiensis*)

Magdalena caiman or Central American caiman (*Caiman crocodilus fuscus*)

Paraguay caiman or Yacare (*Caiman crocodilus yacare*)

Broad-nosed caiman (*Caiman latirostris*)

Black caiman (*Melanosuchus niger*)

American crocodile (*Crocodylus acutus*)

African slender-snouted crocodile (*Crocodylus cataphractus*)

Dwarf or Cuvier's smooth-fronted caiman (*Paleosuchus palpebrosus*)

Smooth-fronted or Schneider's caiman (*Paleosuchus trigonatus*)

Australian freshwater crocodile (*Crocodylus johnsoni*)

New Guinean crocodile (*Crocodylus novaeguineae novaeguineae*)

Ceylon swamp crocodile or Kimbula (*Crocodylus palustris kimbula*)

Estuarine or saltwater crocodile (*Crocodylus porosus*)

	ENDANGERED	VULNERABLE	RARE
	Orinoco crocodile (*Crocodylus intermedius*)		
	Morelet's crocodile (*Crocodylus moreletii*)		
	Nile crocodile (*Crocodylus niloticus*)		
	Mugger, marsh crocodile (*Crocodylus palustris palustris*)		
	Cuban crocodile (*Crocodylus rhombifer*)		
	Siamese crocodile (*Crocodylus siamensis*)		
	Dwarf crocodile (*Osteolaemus tetraspis*)		
	False gavial (*Tomistoma schlegelii*)		
	Gharial or Indian gavial (*Gavialis gangeticus*)		
Squamata	Rodrigues day gecko (*Phelsuma edwardnewtoni*)	Reticulated velvet gecko (*Oedura reticulata*)	Serpent Island gecko (*Cyrtodactylus serpensinsula*)

River terrapin, Tuntong (*Batagur baska*)

South American red-lined turtle (*Pseudemys ornata callirostris*)

South Albemarle tortoise (*Testudo elephantopus elephantopus*)

Abingdon saddlebacked tortoise (*Testudo elephantopus abingdonii*)

North Albemarle saddleback tortoise (*Testudo elephantopus becki*)

Chatham Islands tortoise (*Testudo elephantopus chathamensis*)

James Island tortoise (*Testudo elephantopus darwini*)

Duncan saddlebacked tortoise (*Testudo elephantopus ephippium*)

Vilamil Mountain or Southwest Albemarle tortoise (*Testudo elephantopus guentheri*)

Sail-fin or Soa-soa water lizard (*Hydrosaurus pustulatus*)

Bog turtle (*Clemmys muhlenbergii*)

Aquatic box turtle (*Terrapene coahuila*)

Mediterranean spur-thighed tortoise (*Testudo graeca graeca*)

Radiated tortoise (*Testudo radiata*)

Madagascar tortoise (*Testudo yniphora*)

Loggerhead turtle (*Caretta caretta*)

East Pacific green turtle (*Chelonia mydas agassizi*)

Terecay turtle (*Podocnemis unifilis*)

Round Island day gecko (*Phelsuma guentheri*)

Island night lizard (*Klauberina riversiana*)

Inagua Island turtle (*Chrysemys malonei*)

Desert tortoise, Western gopher tortoise (*Gopherus polyphemus agassizii*)

Berlandier's gopher tortoise, Texas tortoise (*Gopherus polyphemus berlandieri*)

Madagascar spider tortoise (*Pyxis arachnoides*)

Geometric tortoise (*Testudo geometrica*)

Argentine land tortoise (*Geochelone chilensis*)

Flatback green turtle (*Chelonia depressa*)

ENDANGERED	VULNERABLE	RARE
Hood saddlebacked tortoise (*Testudo elephantopus hoodensis*)	Galápagos land iguana (*Conolophus subcristatus*)	Galápagos marine iguana (*Amblyrhynchus cristatus*)
Tagus Cove tortoise (*Testudo elephantopus microphyes*)	Mona Island rhinoceros or ground iguana (*Cyclura cornuta stejnegeri*)	Barrington land iguana (*Conolophus pallidus*)
Indefatigable Island tortoise, Porter's black tortoise (*Testudo elephantopus nigrita*)	Cuban ground iguana (*Cyclura nubila*)	Andros Island ground iguana (*Cyclura baeolopha*)
Cowley Mountain tortoise (*Testudo elephantopus vandenburghi*)	Orange-throated whiptail or race-runner (*Cnemidophorus hyperythrus*)	Turks and Caicos ground iguana (*Cyclura carinata carinata*)
Green turtle (*Chelonia mydas*)	Gila monster (*Heloderma suspectum*)	Mayaguana or Bartsch's rock iguana (*Cyclura carinata bartschi*)
Hawksbill turtle (*Eretmochelys imbricata*)	Central Asian grey monitor (*Varanus griseus caspius*)	Andros Island ground iguana (*Cyclura cychlura cychlura*)
Atlantic ridley turtle (*Lepidochelys kempi*)	Jamaica boa (*Epicrates subflavus*)	Exuma Island ground or rock iguana (*Cyclura cychlura figginsi*)
Olive turtle or Pacific ridley (*Lepidochelys olivacea*)	Indian python (*Python molurus*)	Allen Cays iguana or rock iguana (*Cyclura cychlura inornata*)

Leathery turtle, Luth (*Dermochelys coriacea*)

South American river turtle, Arrau (*Podocnemis expansa*)

Short-necked or swamp turtle (*Pseudemydura umbrina*)

Fiji banded iguana (*Brachylophus fasciatus*)

Blunt-nosed or San Joaquin leopard lizard (*Crotaphytus wislizenii silus*)

Anegada or Virgin Island ground iguana (*Cyclura pinguis*)

Watling Island ground iguana or San Salvador rock iguana (*Cyclura rileyi rileyi*)

Giant or Culebra giant anole (*Anolis roosevelti*)

Hierro giant lizard (*Gallotia simonyi*)

St. Croix ground lizard (*Ameiva polops*)

Black or California legless lizard (*Aniella pulchra nigra*)

Eastern Indigo snake (*Drymarchon corais couperi*)

Fiji snake, Ogmodon (*Ogmodon vitianus*)

Schweizer's Lebetina viper (*Vipera lebetina schweizeri*)

Ridge-nosed rattlesnake (*Crotalus willardi*)

White Cay ground or rock iguana (*Cyclura rileyi cristata*)

Acklins ground or rock iguana (*Cyclura rileyi nuchalis*)

Filfola lizard (*Lacerta filfolensis filfolensis*)

Macabé Forest skink (*Gongylomorphus bojerii fontenayi*)

Round Island skink (*Leiolopisma telfairii*)

Komodo dragon, Komodo monitor or Ora (*Varanus komodoensis*)

Mona blind snake (*Typhlops monensis*)

Southern rubber boa (*Charina bottae umbratica*)

Mona boa (*Epicrates monensis monensis*)

Bimini boa (*Epicrates striatus fosteri*)

San Joaquin whipsnake (*Masticophis flagellum ruddocki*)

ENDANGERED	VULNERABLE	RARE
Round Island boa (*Bolyeria multocarinata*)		Alameda striped racer (*Masticophis lateralis euryxanthus*)
Round Island keel-scaled boa (*Casarea dussumieri*)		Giant garter snake (*Thamnophis couchi gigas*)
Puerto Rican boa or Culebra grande (*Epicrates inornatus*)		Two-striped garter snake (*Thamnophis couchi hammondi*)
San Francisco garter snake (*Thamnophis sirtalis tetrataenia*)		Atlantic saltmarsh snake (*Nerodia fasciata taeniata*)
Central Asian cobra (*Naja oxiana*)		Transcaucasian long-nosed viper (*Vipera ammodytes transcaucasiana*)
Latifi's viper (*Vipera latifi*)		Armenian viper (*Vipera xanthina raddei*)
New Mexico ridge-nosed rattlesnake (*Crotalus willardi obscurus*)		Aruba Island rattlesnake (*Crotalus unicolor*)

BIRDS

Podicipediformes

| | Madagascar red-necked grebe (*Tachybaptus rufolavatus*) | Hooded grebe (*Podiceps gaillardoi*) |

Procellariiformes	Short-tailed albatross (*Diomedea albatrus*)	Westland black petrel (*Procellaria westlandica*)	Gould's petrel (*Pterodroma leucoptera leucoptera*)
	Black petrel (*Procellaria parkinsoni*)	Black-capped petrel or diablotin (*Pterodroma hasitata*)	
	Reunion petrel (*Pterodroma aterrima*)	Newell's shearwater (*Puffinus puffinus newelli*)	
	New Zealand Cook's petrel (*Pterodroma cookii cookii*)		
	Chatham Islands petrel (*Pterodroma hypoleuca axillaris*)		
	Hawaiian dark-rumped petrel (*Pterodroma phaeopygia sandwichensis*)		
Pelecaniformes	Abbott's booby (*Sula abbotti*)	Christmas frigatebird (*Fregata andrewsi*)	
Ciconiiformes	Japanese crested ibis (*Nipponia nippon*)	Chinese or Swinhoe's egret (*Egretta eulophotes*)	Giant ibis (*Thaumatibis gigantea*)
		Milky stork (*Mycteria cinerea*)	Aldabra sacred ibis (*Threskiornis aethiopica abbotti*)
		Madagascar crested ibis (*Lophotibis cristata* subspp.)*	

*Subspecies

	ENDANGERED	VULNERABLE	RARE
Anseriformes	Marianas mallard (*Anas "oustaleti"*)	New Zealand brown teal (*Anas aucklandica* subspp.) Madagascar teal (*Anas bernieri*) Koloa or Hawaiian duck (*Anas platyrhynchos wyvilliana*) Madagascar pochard (*Aythya innotata*) Nene or Hawaiian goose (*Branta sandvicensis*) White-winged wood duck (*Cairina scutulata*) West Indian whistling duck (*Dendrocygna arborea*)	Laysan duck (*Anas laysanensis*)
Falconiformes	Anjouan sparrowhawk (*Accipiter francesii pusillus*) Grenada hook-billed kite (*Chondrohierax uncinatus mirus*) Madagascar serpent-eagle (*Eutriorchis astur*)		Christmas brown goshawk (*Accipiter fasciatus natalis*) Cuban sharp-shinned hawk (*Accipiter striatus fringilloides*) Puerto Rican sharp-shinned hawk (*Accipiter striatus venator*)

Monkey-eating eagle (Philippine eagle) (*Pithecophaga jefferyi*)

Mauritius kestrel (*Falco punctatus*)

Puerto Rican broad-winged hawk (*Buteo platypterus brunnescens*)

Hawaiian hawk or 'Io (*Buteo solitarius*)

Cuban hook-billed kite (*Chondrohierax uncinatus wilsonii*)

Seychelles kestrel (*Falco araea*)

Aldabra kestrel (*Falco newtoni aldabranus*)

Iwo peregrine falcon (*Falco peregrinus fruitii*)

Galliformes

Trinidad piping guan (*Pipile pipile pipile*)

White-eared pheasant (*Crossoptilon crossoptilon*)

Sclater's monal (*Lophophorus sclateri*)

Cheer pheasant (*Catreus wallichii*)

Bulwer's wattled pheasant (*Lophura bulweri*)

Marianas megapode (*Megapodius laperouse laperouse*)

Brown-eared pheasant (*Crossoptilon mantchuricum*)

Imperial pheasant (*Lophura imperialis*)

Palau megapode (*Megapodius laperouse senex*)

Chinese monal (*Lophophorus lhuysii*)

Swinhoe's pheasant (*Lophura swinhoii*)

Hume's bar-tailed pheasant (*Syrmaticus humiae* subspp.)

Elliot's pheasant (*Syrmaticus ellioti*)

Palawan peacock pheasant (*Polyplectron emphanum*)

Blyth's tragopan (*Tragopan blythii* subspp.)

	ENDANGERED	VULNERABLE	RARE
	Cabot's tragopan (*Tragopan caboti*)	Mikado pheasant (*Syrmaticus mikado*)	
Gruiformes	Hawaiian gallinule (*Gallinula chloropus sandvicensis*)	Hooded crane (*Grus monacha*)	Brown mesite (*Mesoenas unicolor*)
	Takahe (*Notornis mantelli*)	Guam rail (*Rallus owstoni*)	White-breasted mesite (*Mesoenas variegata*)
	Barred-wing rail (*Rallus poecilopterus*)		Bensch's monias (*Monias benschi*)
	Lord Howe wood rail (*Tricholimnas sylvestris*)		Cuban sandhill crane (*Grus canadensis nesiotes*)
	Kagu (*Rhynochetos jubatus*)		Zapata rail (*Cyanolimnas cerverai*)
	Great Indian bustard (*Choriotis nigriceps*)		Aldabra white-throated rail (*Dryolimnas cuvieri aldabranus*)
			Hawaiian coot (*Fulica americana alai*)
			Marianas gallinule (*Gallinula chloropus guami*)
Charadriiformes	Chatham Islands oystercatcher (*Haematopus chathamensis*)	Tuamotu sandpiper (*Prosobonia cancellatus*)	New Zealand snipe (*Coenocorypha aucklandica* subspp.)

	New Zealand shore plover (*Thinornis novaeseelandiae*)		Asian dowitcher (*Limnodromus semipalmatus*)
	Black stilt (*Himantopus novaeseelandiae*)		Hawaiian stilt (*Himantopus himantopus knudseni*)
			Relict gull (*Larus relictus*)
Columbiformes	Palau nicobar pigeon (*Caloenas nicobarica pelewensis*)	Tooth-billed pigeon (*Didunculus strigirostris*)	Rapa fruit dove (*Ptilinopus huttoni*)
	Puerto Rican plain pigeon (*Columba inornata wetmorei*)	Cloven-feathered dove (*Drepanoptila holosericea*)	Moheli green pigeon (*Treron australis griveaudi*)
	Marquesas pigeon (*Ducula galeata*)	Society Islands pigeon (*Ducula aurorae*)	
	Truk Micronesian pigeon (*Ducula oceanica teraokai*)	Giant imperial pigeon (*Ducula goliath*)	
	Chatham Islands pigeon (*Hemiphaga novaeseelandiae chathamensis*)	Christmas imperial pigeon (*Ducula whartoni*)	
	Pink pigeon (*Nesoenas mayeri*)	Marianas fruit dove (*Ptilinopus roseicapillus*)	
	Seychelles turtle dove (*Streptopelia picturata rostrata*)		
Psittaciformes	Red-necked parrot or jacquot (*Amazona arausiaca*)	Eastern ground parrot (*Pezoporus wallicus wallicus*)	Bahamas parrot (*Amazona leucocephala bahamensis*)

ENDANGERED	VULNERABLE	RARE
St. Vincent parrot (*Amazona guildingii*)	Thick-billed parrot (*Rhynchopsitta pachyrhyncha pachyrhyncha*)	Orange-bellied parakeet (*Neophema chrysogaster*)
Imperial parrot or sisserou (*Amazona imperialis*)		Splendid or scarlet-chested parakeet (*Neophema splendida*)
St. Lucia parrot (*Amazona versicolor*)		Golden-shouldered parakeet (*Psephotus chrysopterygius chrysopterygius*)
Puerto Rican parrot (*Amazona vittata*)		Hooded parakeet (*Psephotus chrysopterygius dissimilis*)
Seychelles lesser vasa parrot (*Coracopsis nigra barklyi*)		Tahiti lorikeet (*Vini peruviana*)
Chatham Islands yellow-crowned parakeet (*Cyanoramphus auriceps forbesi*)		Ultramarine lorikeet (*Vini ultramarina*)
Orange-fronted parakeet (*Cyanoramphus malherbi*)		
Norfolk Island parakeet (*Cyanoramphus novaeseelandiae cookii*)		
Uvea horned parrot (*Eunymphicus cornutus uvaeensis*)		

... on Prayer from
Oswald Chambers
"My Utmost for His Highest"

THINKING OF PRAYER
AS JESUS TAUGHT

"Pray without ceasing . . ." (1 Thessalonians 5:17).

Our thinking about prayer, whether right or wrong, is based on our own mental conception of it. The correct concept is to think of prayer as the breath in our lungs and the blood from our hearts. Our blood flows and our breathing continues "without ceasing"; we are not even conscious of it, but it never stops. And we are not always conscious of Jesus keeping us in perfect oneness with God, but if we are obeying Him, He always is. Prayer is not an exercise, it is the life of the saint. Beware of anything that stops the offering up of prayer. "Pray without ceasing . . ."—maintain the childlike habit of offering up prayer in your heart to God all the time.

Jesus never mentioned unanswered prayer. He had the unlimited certainty of knowing that prayer is always answered. Do we have through the Spirit of God that inexpressible certainty that Jesus had about prayer, or do we think of the times when it seemed that God did not answer our prayer? Jesus said, ". . . everyone who asks receives . . ." (Matthew 7:8). Yet we say, "But . . . , but . . . ." God answers prayer in the best way—not just sometimes, but every time. However, the evidence of the answer in the area we want it may not always immediately follow. Do we expect God to answer prayer?

The danger we have is that we want to water down what Jesus said to make it mean something that aligns with our common sense. But if it were only common sense, what He said would not even be worthwhile. The things Jesus taught about prayer are supernatural truths He reveals to us.

MAY 26

HOLINESS OR HARDNESS
TOWARD GOD?

"He . . . wondered that there was no intercessor . . ." (Isaiah 59:16).

The reason many of us stop praying and become
hard toward God is that we only have an emotional interest in prayer. It sounds good to say
that we pray, and we read books on prayer which tell us
that prayer is beneficial—that our minds are quieted
and our souls are uplifted when we pray. But Isaiah
implied in this verse that God is amazed at such
thoughts about prayer.

Worship and intercession must go together; one is
impossible without the other. Intercession means raising ourselves up to the point of getting the mind of
Christ regarding the person for whom we are praying
(see Philippians 2:5). Instead of worshiping God, we
recite speeches to God about how prayer is supposed to
work. Are we worshiping God or disputing Him when
we say, "But God, I just don't see how you are going to
do this"? This is a sure sign that we are not worshiping.
When we lose sight of God, we become hard and dogmatic. We throw our petitions at His throne and dictate to Him what we want Him to do. We don't
worship God, nor do we seek to conform our minds to
the mind of Christ. And if we are hard toward God, we
will become hard toward other people.

Are we worshiping God in a way that will raise us up
to where we can take hold of Him, having such intimate
contact with Him that we know His mind about the
ones for whom we pray? Are we living in a holy relationship with God, or have we become hard and dogmatic?

Do you find yourself thinking that there is no one
interceding properly? Then be that person yourself. Be
a person who worships God and lives in a holy relationship with Him. Get involved in the real work of
intercession, remembering that it truly is work—work
that demands all your energy, but work which has no
hidden pitfalls. Preaching the gospel has its share of
pitfalls, but intercessory prayer has none whatsoever.

MARCH 30

THE PURPOSE OF PRAYER

". . . one of His disciples said to Him, 'Lord, teach us to pray . . .' " (Luke 11:1).

Prayer is not a normal part of the life of the natural man. We hear it said that a person's life will suffer if he doesn't pray, but I question that. What will suffer is the life of the Son of God in him, which is nourished not by food, but by prayer. When a person is born again from above, the life of the Son of God is born in him, and he can either starve or nourish that life. Prayer is the way that the life of God in us is nourished. Our common ideas regarding prayer are not found in the New Testament. We look upon prayer simply as a means of getting things for ourselves, but the biblical purpose of prayer is that we may get to know God Himself.

"Ask, and you will receive . . ." (John 16:24). We complain before God, and sometimes we are apologetic or indifferent to Him, but we actually *ask* Him for very few things. Yet a child exhibits a magnificent boldness to ask! Our Lord said, ". . . unless you . . . become as little children . . ." (Matthew 18:3). Ask and God will do. Give Jesus Christ the opportunity and the room to work. The problem is that no one will ever do this until he is at his wits' end. When a person is at his wits' end, it no longer seems to be a cowardly thing to pray; in fact, it is the only way he can get in touch with the truth and the reality of God Himself. Be yourself before God and present Him with your problems—the very things that have brought you to your wits' end. But as long as you think you are self-sufficient, you do not need to ask God for anything.

To say that "prayer changes things" is not as close to the truth as saying, "Prayer changes me and then I change things." God has established things so that prayer, on the basis of redemption, changes the way a person looks at things. Prayer is not a matter of changing things externally, but one of working miracles in a person's inner nature.

PRAYER IN THE FATHER'S HEARING

"Jesus lifted up His eyes and said, 'Father, I thank You that You have heard Me' " (John 11:41).

When the Son of God prays, He is only mindful and consciously aware of His Father. God always hears the prayers of His Son, and if the Son of God has been formed in me (see Galatians 4:19) the Father will always hear my prayers. But I must see to it that the Son of God is exhibited in my human flesh. ". . . your body is the temple of the Holy Spirit . . ." (1 Corinthians 6:19) that is, your body is the Bethlehem of God's Son. Is the Son of God being given His opportunity to work in me? Is the direct simplicity of His life being worked out in me exactly as it was worked out in His life while here on earth? When I come into contact with the everyday occurrences of life as an ordinary human being, is the prayer of God's eternal Son to His Father being prayed in me? Jesus says, "In that day you will ask in My name . . ." (John 16:26). What day does He mean? He is referring to the day when the Holy Spirit has come to me and made me one with my Lord.

Is the Lord Jesus Christ being abundantly satisfied by your life, or are you exhibiting a walk of spiritual pride before Him? Never let your common sense become so prominent and forceful that it pushes the Son of God to one side. Common sense is a gift that God gave to our human nature—but common sense is not the gift of His Son. Supernatural sense is the gift of His Son, and we should never put our common sense on the throne. The Son always recognizes and identifies with the Father, but common sense has never yet done so and never will. Our ordinary abilities will never worship God unless they are transformed by the indwelling Son of God. We must make sure that our human flesh is kept in perfect submission to Him, allowing Him to work through it moment by moment. Are we living at such a level of human dependence upon Jesus Christ that His life is being exhibited moment by moment in us?

AUGUST 9

THE EXPLANATION FOR OUR DIFFICULTIES

"... that they all may be one, as You, Father, are in Me, and I in You; that they also may be one in Us ..." (John 17:21).

If you are going through a time of isolation, seemingly all alone, read John 17. It will explain exactly why you are where you are—because Jesus has prayed that you "may be one" with the Father as He is. Are you helping God to answer that prayer, or do you have some other goal for your life? Since you became a disciple, you cannot be as independent as you used to be.

God reveals in John 17 that His purpose is not just to answer our prayers, but that through prayer we might come to discern His mind. Yet there is one prayer which God must answer, and that is the prayer of Jesus—"... that they may be one just as We are one ..." (17:22). Are we as close to Jesus Christ as that?

God is not concerned about our plans; He doesn't ask, "Do you want to go through this loss of a loved one, this difficulty, or this defeat?" No, He allows these things for His own purpose. The things we are going through are either making us sweeter, better, and nobler men and women, or they are making us more critical and fault-finding, and more insistent on our own way. The things that happen either make us evil, or they make us more saintly, depending entirely on our relationship with God and its level of intimacy. If we will pray, regarding our own lives, "Your will be done" (Matthew 26:42), then we will be encouraged and comforted by John 17, knowing that our Father is working according to His own wisdom, accomplishing what is best. When we understand God's purpose, we will not become small-minded and cynical. Jesus prayed nothing less for us than absolute oneness with Himself, just as He was one with the Father. Some of us are far from this oneness; yet God will not leave us alone until we are one with Him—because Jesus prayed, "... that they all may be one ...."

MAY 22

PRAYER—BATTLE IN "THE SECRET PLACE"

"When you pray, go into your room, and when you have shut your door, pray to your Father who is in the secret place; and your Father who sees in secret will reward you openly" (Matthew 6:6).

Jesus did not say, "Dream about your Father who is in the secret place," but He said, ". . . pray to your Father who is in the secret place. . . ." Prayer is an effort of the will. After we have entered our secret place and shut the door, the most difficult thing to do is to pray. We cannot seem to get our minds into good working order, and the first thing we have to fight is wandering thoughts. The great battle in private prayer is overcoming this problem of our idle and wandering thinking. We have to learn to discipline our minds and concentrate on willful, deliberate prayer.

We must have a specially selected place for prayer, but once we get there this plague of wandering thoughts begins, as we begin to think to ourselves, "This needs to be done, and I have to do that today." Jesus says to "shut your door." Having a secret stillness before God means deliberately shutting the door on our emotions and remembering Him. God is in secret, and He sees us from "the secret place"—He does not see us as other people do, or as we see ourselves. When we truly live in "the secret place," it becomes impossible for us to doubt God. We become more sure of Him than of anyone or anything else. Enter into "the secret place," and you will find that God was right in the middle of your everyday circumstances all the time. Get into the habit of dealing with God about everything. Unless you learn to open the door of your life completely and let God in from your first waking moment of each new day, you will be working on the wrong level throughout the day. But if you will swing the door of your life fully open and "pray to your Father who is in the secret place," every public thing in your life will be marked with the lasting imprint of the presence of God.

AUGUST 23

PRAYING TO GOD IN SECRET

"When you pray, go into your room, and when you have shut your door, pray to your Father who is in the secret place . . ." (Matthew 6:6).

The primary thought in the area of religion is—keep your eyes on God, not on people. Your motivation should not be the desire to be known as a praying person. Find an inner room in which to pray where no one even knows you are praying, shut the door, and talk to God in secret. Have no motivation other than to know your Father in heaven. It is impossible to carry on your life as a disciple without definite times of secret prayer.

"When you pray, do not use vain repetitions . . ." (6:7). God does not hear us because we pray earnestly—He hears us solely on the basis of redemption. God is never impressed by our earnestness. Prayer is not simply getting things from God—that is only the most elementary kind of prayer. Prayer is coming into perfect fellowship and oneness with God. If the Son of God has been formed in us through regeneration (see Galatians 4:19), then He will continue to press on beyond our common sense and will change our attitude about the things for which we pray.

"Everyone who asks receives . . ." (Matthew 7:8). We pray religious nonsense without even involving our will, and then we say that God did not answer—but in reality we have never *asked* for anything. Jesus said, ". . . you will ask what you desire . . ." (John 15:7). Asking means that our will must be involved. Whenever Jesus talked about prayer, He spoke with wonderful childlike simplicity. Then we respond with our critical attitude, saying, "Yes, but even Jesus said that we must *ask*." But remember that we have to ask things of God that are in keeping with the God whom Jesus Christ revealed.

SEPTEMBER 16

THE UNRIVALED
POWER OF PRAYER

"We do not know what we should pray for as we ought, but the Spirit Himself makes intercession for us with groanings which cannot be uttered" (Romans 8:26).

We realize that we are energized by the Holy Spirit for prayer; and we know what it is to pray in accordance with the Spirit; but we don't often realize that the Holy Spirit Himself prays prayers in us which we cannot utter ourselves. When we are born again of God and are indwelt by the Spirit of God, He expresses for us the unutterable.

"He," the Holy Spirit in you, "makes intercession for the saints according to the will of God" (8:27). And God searches your heart, not to know what your conscious prayers are, but to find out what the prayer of the Holy Spirit is.

The Spirit of God uses the nature of the believer as a temple in which to offer His prayers of intercession. ". . . your body is the temple of the Holy Spirit . . ." (1 Corinthians 6:19). When Jesus Christ cleansed the temple, ". . . He would not allow anyone to carry wares through the temple" (Mark 11:16). The Spirit of God will not allow you to use your body for your own convenience. Jesus ruthlessly cast out everyone who bought and sold in the temple, and said, "My house shall be called a house of prayer . . . . But you have made it a 'den of thieves' " (Mark 11:17).

Have we come to realize that our "body is the temple of the Holy Spirit"? If so, we must be careful to keep it undefiled for Him. We have to remember that our conscious life, even though only a small part of our total person, is to be regarded by us as a "temple of the Holy Spirit." He will be responsible for the unconscious part which we don't know, but we must pay careful attention to and guard the conscious part for which we are responsible.

NOVEMBER 8

INTERCESSORY PRAYER

" . . . men always ought to pray and not lose heart" (Luke 18:1).

You cannot truly intercede through prayer if you do not believe in the reality of redemption. Instead, you will simply be turning intercession into useless sympathy for others, which will serve only to increase the contentment they have for remaining out of touch with God. True intercession involves bringing the person, or the circumstance that seems to be crashing in on you, before God, until you are changed by His attitude toward that person or circumstance. Intercession means to "fill up . . . [with] what is lacking in the afflictions of Christ" (Colossians 1:24), and this is precisely why there are so few intercessors. People describe intercession by saying, "It is putting yourself in someone else's place." That is not true! Intercession is putting yourself in God's place; it is having His mind and His perspective.

As an intercessor, be careful not to seek too much information from God regarding the situation you are praying about, because you may be overwhelmed. If you know too much, more than God has ordained for you to know, you can't pray; the circumstances of the people become so overpowering that you are no longer able to get to the underlying truth.

Our work is to be in such close contact with God that we may have His mind about everything, but we shirk that responsibility by substituting doing for interceding. And yet intercession is the only thing that has no drawbacks, because it keeps our relationship completely open with God.

What we must avoid in intercession is praying for someone to be simply "patched up." We must pray that person completely through into contact with the very life of God. Think of the number of people God has brought across our path, only to see us drop them! When we pray on the basis of redemption, God creates something He can create in no other way than through intercessory prayer.

DECEMBER 13

Western ground parrot
(*Pezoporus wallicus
flaviventris*)

Paradise or Beautiful parakeet
(*Psephotus pulcherrimus*)

Mauritius parakeet
(*Psittacula echo*)

Maroon-fronted parrot
(*Rhynchopsitta
pachyrhyncha terrisi*)

Kakapo (*Strigops
habroptilus*)

Strigiformes

Soumagne's Owl (*Tyto
soumagnei*)

Anjouan scops owl (*Otus
rutilus capnodes*)

Tobago striped owl (*Asio
clamator oberi*)

Ponape short-eared owl (*Asio
flammeus ponapensis*)

Christmas Island owl (*Ninox
squamipila natalis*)

Seychelles owl (*Otus
insularis*)

Virgin Island screech owl
(*Otus nudipes newtoni*)

Caprimulgiformes

Puerto Rican Whippoorwill
(*Caprimulgus noctitherus*)

	ENDANGERED	VULNERABLE	RARE
Coraciiformes	Guam Micronesian kingfisher (*Halcyon cinnamomina cinnamomina*)	Long-tailed ground roller (*Uratelornis chimaera*)	Crossley's ground roller (*Atelornis crossleyi*) Pitta-like ground roller (*Atelornis pittoides*) Short-legged ground roller (*Brachypteracias leptosomus*) Scaled ground roller (*Brachypteracias squamigera*)
Piciformes	Cuban ivory-billed woodpecker (*Campephilus principalis bairdii*) Tristram's woodpecker (*Dryocopus javensis richardsi*) Okinawa woodpecker (*Sapheopipo noguchii*)		Owston's white-backed woodpecker (*Dendrocopos leucotos owstoni*) Inouye's three-toed woodpecker (*Picoides tridactylus inouyei*) Takatsukasa's green woodpecker (*Picus awokera takatsukasae*)
Passeriformes	St. Lucia forest thrush (*Cichlherminia lherminieri sanctaeluciae*)	Hawaii 'akepa (*Loxops coccinea coccinea*)	St. Vincent solitaire (*Myadestes genibarbis sibilans*)

Seychelles magpie-robin
(*Copsychus sechellarum*)

Southern Ryukyu robin
(*Erithacus komadori
subrufa*)

Kauai thrush (*Phaeornis
obscurus myadestina*)

Molokai thrush (*Phaeornis
obscurus rutha*)

Puaiohi or small kauai thrush
(*Phaeornis palmeri*)

Grey-headed blackbird
(*Turdus poliocephalus
poliocephalus*)

Eiao Polynesian warbler
(*Acrocephalus caffer
aquilonis*)

Moorea Polynesian warbler
(*Acrocephalus caffer
longirostris*)

Rodrigues brush warbler
(*Bebrornis rodericana*)

Long-legged warbler
(*Trichocichla rufa subspp.*)

Crested honeycreeper
(*Palmeria dolei*)

Maui parrotbill (*Pseudonestor
xanthophrys*)

North Island kokako
(*Callaeas cinerea wilsoni*)

Hatutu Polynesian warbler
(*Acrocephalus caffer
postremus*)

Nihoa millerbird
(*Acrocephalus familiaris
kingi*)

Japanese marsh warbler
(*Megalurus pryeri pryeri*)

Po'o uli (*Melamprosops
phaeosoma*)

Kauai creeper (*Paroreomyza
Loxops maculata bairdi*)

ENDANGERED	VULNERABLE	RARE
Maui nukupu'u (*Hemignathus lucidus affinis*)		
Kauai nukupu'u (*Hemignathus lucidus hanapepe*)		
Kauai 'akialoa (*Hemignathus obscurus procerus*)		
'Akiapola'au (*Hemignathus wilsoni*)		
Palila (*Loxioides Psittirostra bailleui*)		
Maui 'akepa (*Loxops coccinea ochracea*)		
Molokai creeper (*Paroreomyza Loxops maculata flammea*)		
Oahu creeper (*Paroreomyza Loxops maculata maculata*)		
'O'u (*Psittirostra psittacea*)		
South Island kokako (*Callaeas cinerea cinerea*)		
Marianas crow (*Corvus kubaryi*)		
Hawaiian crow (*Corvus tropicus*)		

Index